A NEW COURSE OF PLANTS AND ANIMALS

BOOK II

A NEW COURSE OF PLANTS AND ANIMALS

BY

M. A. GRIGG, B.Sc.

Senior Biology Mistress, Ealing Grammar School for Girls
Formerly Lecturer in Biology, Dudley Training College for Teachers

BOOK II

CAMBRIDGE
AT THE UNIVERSITY PRESS
1963

CAMBRIDGE
UNIVERSITY PRESS

University Printing House, Cambridge CB2 8BS, United Kingdom

Published in the United States of America by Cambridge University Press, New York

Cambridge University Press is part of the University of Cambridge.

It furthers the University's mission by disseminating knowledge in the pursuit of education, learning and research at the highest international levels of excellence.

www.cambridge.org
Information on this title: www.cambridge.org/9781107672000

First published 1958
First edition 1958
Reprinted 1963
First paperback edition 2014

A catalogue record for this publication is available from the British Library

ISBN 978-1-107-67200-0 Paperback

PREFACE

Biology is essentially a study of life, and not a book study. The best lessons are those which can be taken out-of-doors where living things can be studied in their natural surroundings. Unfortunately, very few of us can do this during school hours, so we must bring as many specimens as possible to school for examination, and then study them in their natural surroundings when we can.

In Secondary Grammar Schools, the Biology syllabus is sometimes ruled by the syllabus of the G.C.E. examinations, and general Nature Study is often neglected. The two books in this series are intended to guide the studies of boys and girls in Nature Study during the first three years of their Grammar School life, and to stimulate their interest in the living things around them. When once their interest has been aroused, they will be more observant, and will wish to learn more about the specimens which they find. They may then refer to the books mentioned in Appendix C and so acquire the habit of seeking for information on their own.

The first course of *Plants and Animals* was a re-issue of the Biology chapters from *Elementary Science*, which was written primarily for Secondary Modern Schools. The scientific terms were omitted so that all could read and understand the subject. This new course, based on the Biology chapters of *Modern Science*, has been entirely re-written, and some scientific terms have been introduced which will be useful for pupils who intend to take the G.C.E. examination. Where scientific terms have been used, simple descriptions have also been given so that the book is suitable for all boys and girls in Secondary Modern Schools.

In Book I the work was based on the study of plants and animals found in habitats familiar to most boys and girls. It is impossible to study all the specimens that we find or see; but we can study families, each family consisting of animals or plants which are similar in structure. The first five chapters of Book II have been based on a simple study of families. This work not only incorporates the specimens described in Book I, but also gives an account of inhabitants of the sea shores, and of animals that can be seen in zoological gardens. Many common plants that do not have flowers have also been described. Simple experimental work has been given to enable pupils to study how a plant lives.

A chapter on Human Physiology has been included, as it is felt that all pupils should have some knowledge of the structure and functions of the body.

As far as possible, this book should be read with the living specimens at hand. Reading should guide observation. If animals are to be studied which cannot be brought to school, the teacher should illustrate the lesson by showing pictures (with or without epidiascope), film strips or films. Visits to a zoological garden or a museum would make the work more interesting.

M. A. GRIGG

27 July 1957

ILLUSTRATIONS

The following were drawn by Miss J. B. S. Willmore: 4 (*c*) and (*d*), 9, 10, 11 (*a*), 33 (*c*) and (*d*), 75, 76, 81, 88 and 113. The rest are the work of the late Mr J. C. Hill.

CONTENTS

CHAPTER 1

PLANTS AND ANIMALS

Reason for classification

In Book I we learned a little about some animals and plants that live in different habitats.

Whilst studying the animals that live in fresh water, in the garden, or in the countryside you may have noticed that some animals are similar in structure; for example, the water skater and the wood louse; pond snails, land snails and slugs; gnats and butterflies. Unfortunately, we have not time to study every animal and plant that we find. All animals and plants, however, have been divided into families, and into each family are put those animals and plants which are similar in structure. The families that have certain things in common are placed together in still larger groups.

If we study one animal or plant from each family we shall know something about the structure of each member of that family, although the members or SPECIES of the family will differ in some respects from one another. You probably found it very easy to study land snails and slugs after studying pond snails. Living things are today classified by their structure, but in the past they have been classified in many different ways—according to their size, or where they lived, or whether they did or did not lay eggs.

Differences between plants and animals

All living things can be divided into two groups: (1) PLANTS; (2) ANIMALS.

How can you tell into which group a living thing should be placed? You will probably say that animals move about from

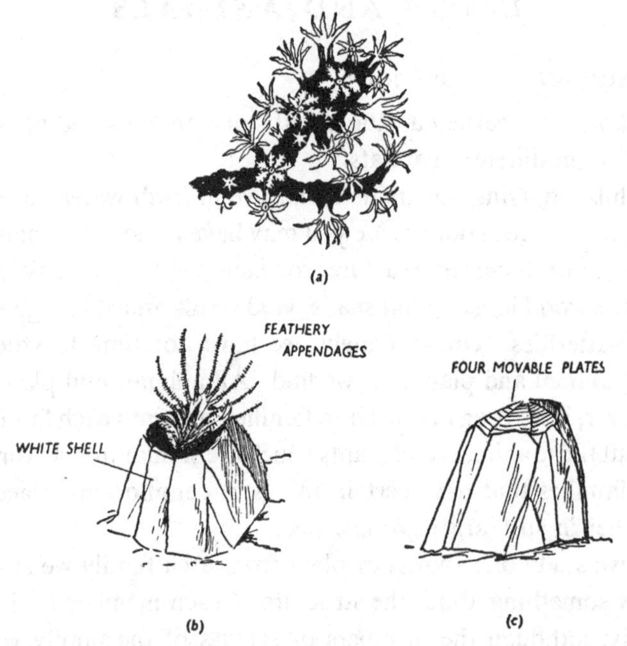

Fig. 1. *Animals that do not move. (a) Coral: the animals are embedded in the calcium carbonate shell; (b) barnacle showing feathery appendages; (c) barnacle with plates closed*

place to place, whilst plants remain rooted in the ground. This is true of most living things, but some animals, such as those which form coral, and barnacles (Figs. 1*a* and *b*), are fixed in one place; whereas some small algae swim about in the water (Fig. 2).

There are six chief differences between animals and plants:

(1) The CELLS of plants are surrounded by a cell wall, which is made of a substance called CELLULOSE. This is not so in animals, where each cell is surrounded by a membrane.

(2) Plants usually contain CHLOROPHYLL, but this is not found in animals. Some plants, such as bacteria and fungi, do not contain chlorophyll. The green hydra, on the other hand, although it is an animal, does contain chlorophyll, but it is not contained in the cells of the hydra's body but in small, round cells, which are really tiny plants that live between the cells of the hydra.

(3) Plants that contain chlorophyll are able to make their own food. During the daytime the green parts of a plant are able to make sugar from carbon dioxide and water. This food, together with the minerals that are taken in by the roots, is changed into more complicated substances. Animals, however, cannot make their own food. They eat the foods that have been made by the plants, either by eating the plants or by eating animals that feed on plants.

(4) Animals usually move from place to place in search of food, whereas plants remain in one place as they can make their own food.

(5) Animals grow to a limited size and the number of parts in their bodies is fixed. Plants continue to grow as their size is unlimited.

(6) Growth in animals goes on all over the body. In plants there are special growing regions, which are at the tips of the roots or shoots, or just beneath the bark in the stems and the roots.

Groups of living things

Plants can be divided into two large groups: (1) flowerless plants that do not produce seeds; (2) plants that do produce seeds.

Animals also can be divided into two groups: (1) INVERTEBRATES, which have no backbone; (2) VERTEBRATES, which have a backbone.

CHAPTER 2

FLOWERLESS PLANTS

There are many plants that do not have flowers, and which are very different from the plants that we have studied so far. We should know something about these plants, not only because many of them are frequently found, but also because some of them are economically very important. These plants have not any flowers or any seeds, but they produce SPORES from which new plants grow.

The green scum on damp palings or on tree trunks, the green scum on ponds, seaweed, mould, yeast, the orange spots on damp wood, the pink spots on dead wood, mushrooms, toadstools, puff balls, mosses and ferns, and even bacteria are flowerless plants. Try to make a collection of flowerless plants and identify them with the help of the books mentioned in Appendix C.

Algae

In this group there are hundreds of plants which are very simple in structure, and which nearly all live in water. Some of them are very big, although their structure is simple, but many of them are so small that they can only be seen through a microscope (Fig. 2). These very tiny algae swim about in the water by means of hair-like projections which may be called CILIA or FLAGELLA. The green scum that you may have seen on ponds consists of very long, fine filaments of algae.

A very common one is called SPIROGYRA (Fig. 3). Place a single thread on a slide in a drop of water and cover it with a coverslip. Look at it under a microscope and you will see that it is divided into cells. Each cell has a cellulose cell wall which is covered with a slimy substance to prevent other organisms from growing on it. Protoplasm lines the cell wall.

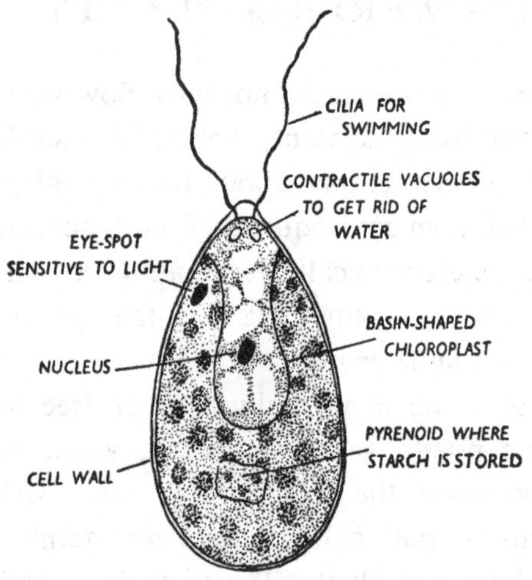

Fig. 2. *Chlamydomonas. A small alga that swims about in water*

The middle of the cell is filled with cell sap and is called a VACUOLE. A NUCLEUS lies in the vacuole, surrounded by protoplasm, threads of which stretch to the cell wall. You will see one or more green spiral bands. These are the SPIRAL CHLOROPLASTS which contain chlorophyll. As the plant contains chlorophyll it makes sugar which can be changed into starch and stored in certain places on the chloroplasts, called the PYRENOIDS.

A strand may break up into pieces and each piece can grow into a new strand. Spirogyra has another way of reproducing itself especially when conditions are not favourable. Two strands come together and lie side by side, glued together by a slimy substance called mucilage. All down the strands lumps grow out from two cells that lie opposite to one another. These lumps eventually join and the strands now look something like a ladder (Fig. 3*b*). The contents of the cells shrink away from the cell wall and round themselves off.

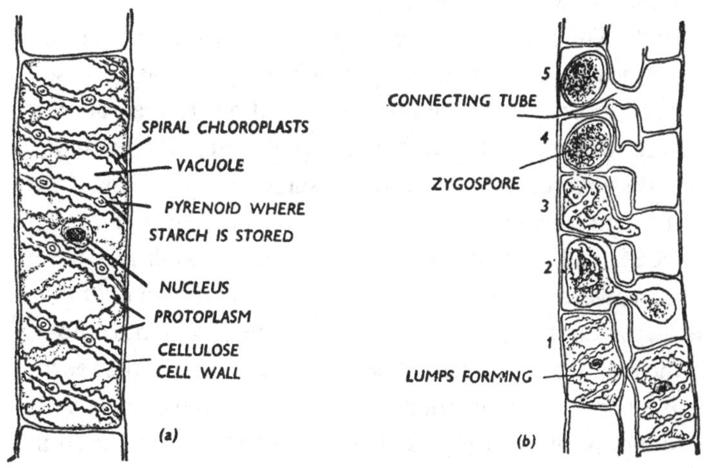

Fig. 3. *Spirogyra. A filamentous alga.* (*a*) *One cell;* (*b*) *reproduction*

Those in the cells of one strand pass along the connecting tubes into the cells of the other strand. Here the two join together and become surrounded by a thick wall. These thick-walled ZYGOSPORES, as they are called, are set free when the cell walls break and fall to the bottom of the pond. They grow into new filaments when conditions are favourable. The old filaments rot away.

Seaweeds

Seaweeds are algae that live in salt water. They vary considerably in size, shape and colour. Make a collection of as many kinds of seaweed as you can find. Some of them will keep when they are dry, other seaweeds must be dried carefully between blotting paper and then glued to paper. Seaweeds are grouped according to their colour, which may be brown, green or red. Although they vary in colour, they all contain chlorophyll and can manufacture their own food. Pigments mask the green chlorophyll in the brown and red seaweeds. The plants are very simple in structure. At one end there is usually a HOLD-FAST by which the plant fastens itself to a rock to prevent it from being washed out to sea. BROWN SEAWEEDS are very common between high and low tide marks. They are covered with a slimy substance which keeps them moist when the tide goes out. One very common brown seaweed is called BLADDERWRACK (Fig. 4*a*). Look at a piece of this seaweed and you will see large and small round lumps. The larger lumps are air sacs which enable the plant to float, but the smaller ones, called CONCEPTACLES contain the reproductive organs. In bladderwrack the male and female cells are found in different plants. The SPERM cells are produced in ANTHERIDIA (Fig. 4*c*) and the EGG cells in OOGONIA (Fig. 4*d*). When the sperms are ripe they escape into the water through the hole or OSTIOLE at the top of the conceptacle. They swim by means of two tails to the egg cells which have escaped into the water. When an egg cell is fertilized a cell wall forms round it. It falls to the bottom of the water, attaches itself to a rock or to a stone, and begins to grow into a new plant.

The serrated wrack (Fig. 4*b*) is similar to the bladderwrack, but it has not any bladders.

GREEN SEAWEEDS grow in places where they are just covered with water. Two common green seaweeds are the sea lettuce and the sea grass (Figs. 5*a* and *b*).

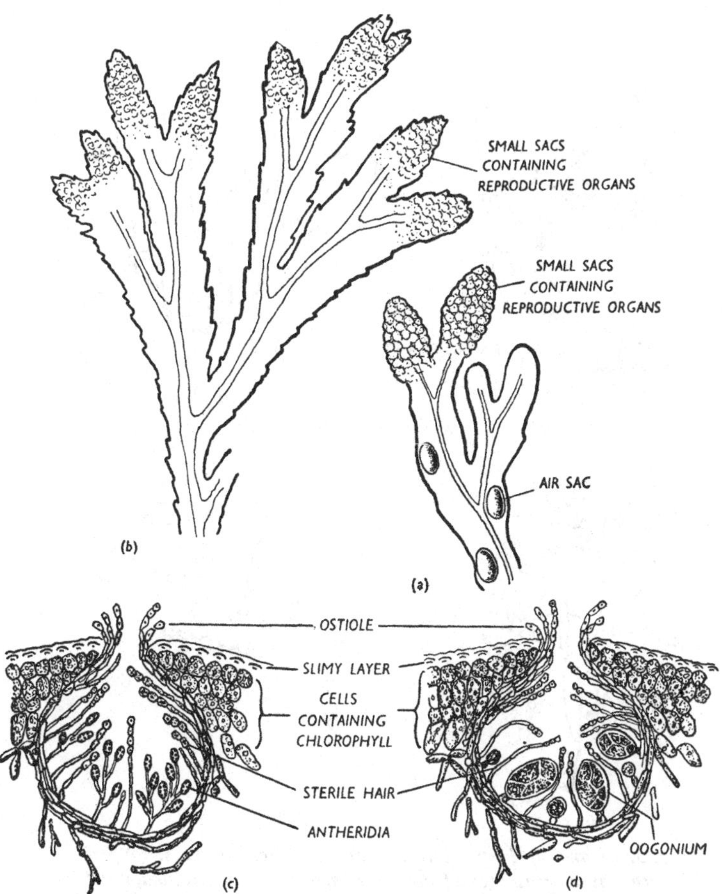

Fig. 4. *Brown seaweeds.* (*a*) *Bladderwrack;* (*b*) *serrated wrack;* (*c*) *section through conceptacle containing antheridia;* (*d*) *section through conceptacle containing oogonia*

RED SEAWEEDS live in deeper water. Some are very delicate and finely divided (Fig. 5*c*), whilst others are covered with a chalky substance that is rather like the shell of a crab (Fig. 5*d*).

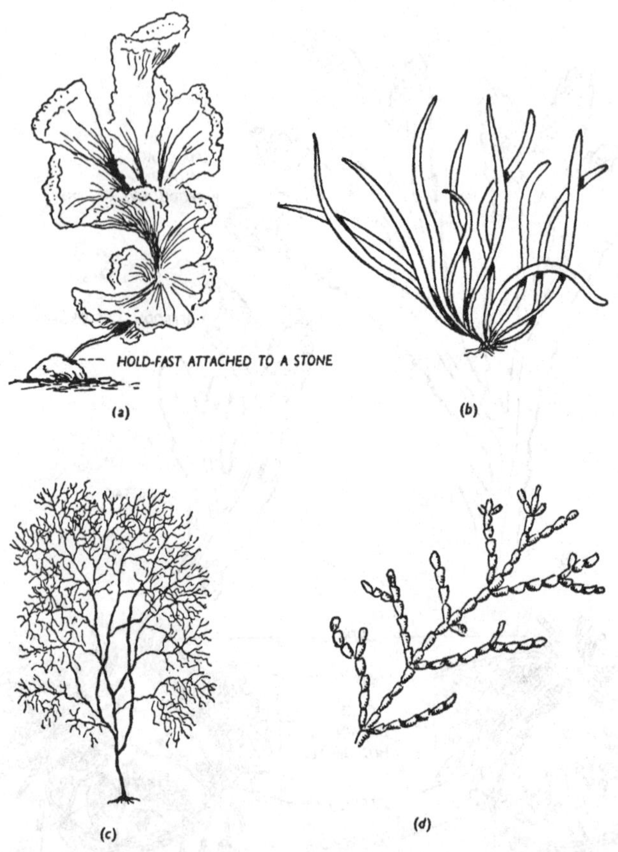

Fig. 5. *(a) and (b) Green seaweeds; (c) and (d) red seaweeds. (a) Sea lettuce; (b) sea grass; (c) a feathery red seaweed; (d) Corallina, a red seaweed which looks like coral*

Fungi

Unlike most plants, fungi do not contain chlorophyll, and so they are not able to make their own food. They get their food from dead things (then they are called SAPROPHYTES) or from living things (they are then called PARASITES). During the autumn look in fields, woods, hedgerows and gardens and you will find a great variety of fungi. Many of them are commonly called TOADSTOOLS, and, although they do not contain chlorophyll, they may be brightly coloured, red, orange, yellow, purple, brown or even green. Toadstools usually live on the humus in the ground, and so they are not harmful. The toadstool itself is not the main part of the plant although it is the biggest part. It is only the reproductive part of a flowerless plant, just as the flower is the reproductive part of a flowering plant. All fungi consist of white threads or HYPHAE. These threads are very narrow, so that they can penetrate either through the decaying matter or through or between the cells of the living animal or plant in which they live. The hyphae give out substances which dissolve the organic matter, changing it into liquids which they can then absorb as food. You can see how dangerous this is if the threads are growing inside a living animal or plant. Try to find some of these hyphae in the soil beneath a toadstool, and look at them under a microscope.

Look at a mushroom (Fig. 6*a*). It consists of a STALK and an umbrella-shaped top or HEAD. On the underside of the mushroom you will see a number of GILLS on which the SPORES are formed (Fig. 6*b*). When the spores are ripe the toadstool may dry and shrivel up, or, as in the ink-cap fungus (Fig. 7*a*), it might dissolve into a liquid which eventually dries up. In both cases the spores, which are very tiny, are blown away. Some toadstools, for example the edible boletus (Fig. 7*b*), which is orange on top and yellow below the head and which is 3 or 4 inches across, and bracket fungus (Fig. 7*c*), which grows like brackets on the trunks of trees, do not have gills on the underside of the head, but they have holes through which the spores can fall. The spores do not always develop on the underside of a special head. In candle snuff fungus

(Fig. 7*d*) the spores grow over the whole of the surface of that part of the fungus which is above the ground. In the earth balls and the puff balls (Fig. 7*e*) the spores are formed inside the 'balls', and they are not released and scattered until the puff ball bursts.

Several varieties of fungus may be eaten, but do not eat them unless you are sure that they are edible, as many fungi

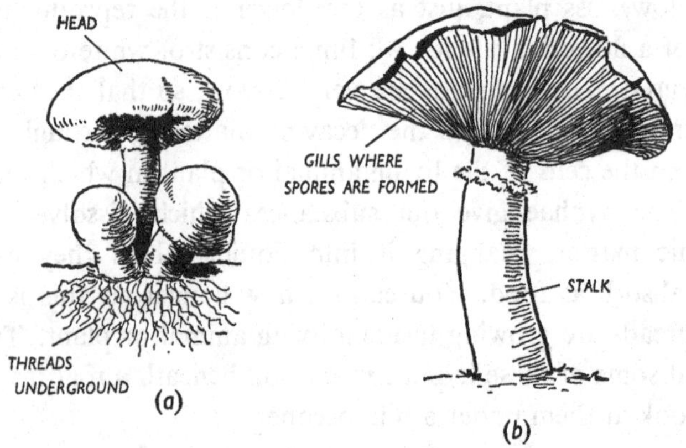

Fig. 6. (*a*) *Young mushrooms;* (*b*) *older mushroom seen from below.*

are poisonous. You can buy books which are entitled *Edible Fungi*, but when using these books to identify specimens, remember that there are many fungi that are similar in colour and it is not easy for a beginner to recognize one from another.

Harmful fungi

All the parasitic fungi are harmful because they live in or on a living animal or plant. The disease called ringworm, which is sometimes found in children, is caused by a fungus which lives on the head or on the face. Thrush is caused by a fungus that grows in the throat. Fish that live in ponds are some-

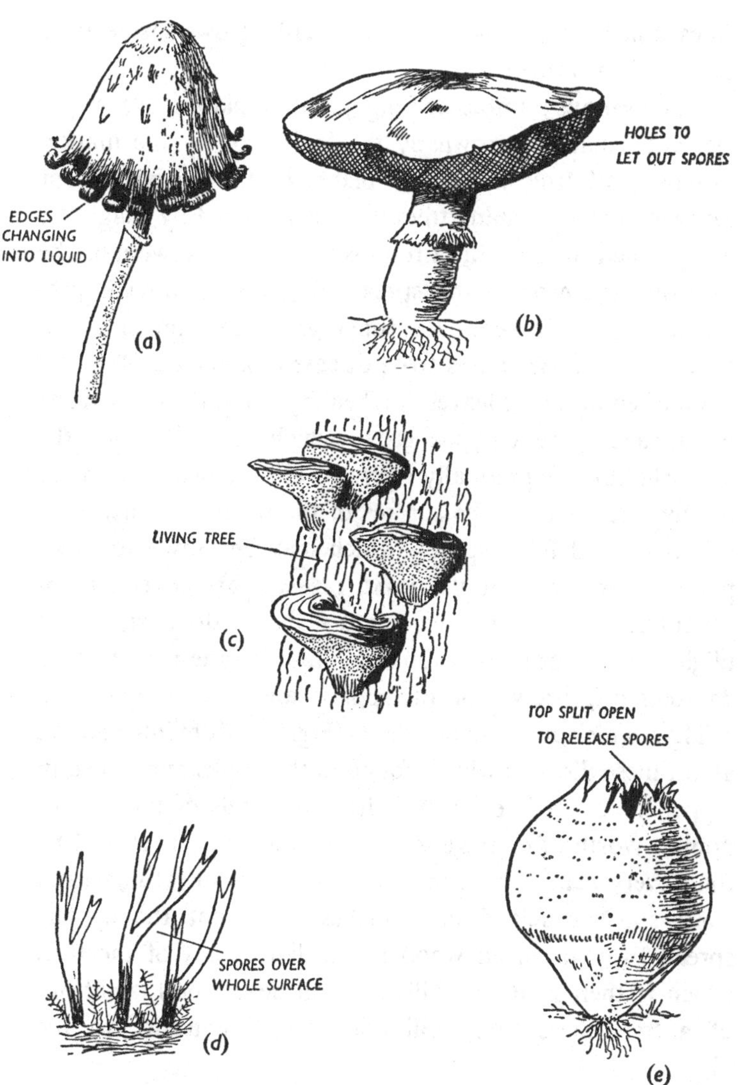

Fig. 7. *Fungi.* (*a*) *Ink cap;* (*b*) *edible boletus;* (*c*) *bracket fungus on silver birch tree;* (*d*) *candle snuff fungus;* (*e*) *puff ball*

times killed by a fungus which actually grows inside their body and kills them.

An enormous number of fungi grow on plants. Nearly all the crops that are grown by the farmer or by the market gardener, all fruit trees, and plants that we grow in our gardens as well as many trees may be attacked by fungi. So today, when the growth of food is so important, we must be sure that the growth and spread of parasitic fungi is prevented or is checked. All fungi grow best under moist conditions. If the spores from diseased plants are blown or are washed on to the leaves of a healthy plant, they will grow. From each spore a hypha grows which penetrates into the leaf. The threads gradually injure or kill the plant. Some of the hyphae come to the surface of the plant on which they are living and form spores, which may be blown to other plants. Some common FUNGAL DISEASES are silver leaf of plum trees, potato dry-rot, potato scab, potato wart, potato blight, club root of cabbage, onion white-rot, tomato damping-off, brown-rot of apples, and rust fungus and mildews which attack many plants (Fig. 8). More information about these diseases will be found in the books mentioned in Appendix C. It is estimated that thousands of millions of pounds worth of damage to crops in the world is caused by fungi every year. Some fungi, for example bracket fungus and coral spot, attack dead branches only, but the hyphae spread from the dead wood to the living part of the tree, which is then gradually killed. If all dead wood is cut out of a tree, these fungi will not gain an entrance to the plant.

The spread of fungi may be prevented by burning all diseased plants. Fungi that live in the soil, for example club root, can be killed by adding lime to the soil. Any spores that are on the leaves and branches can be killed by spraying them with various substances.

Dry-rot of wood is caused by a fungus which gradually softens the wood until it crumbles away to a fine powder.

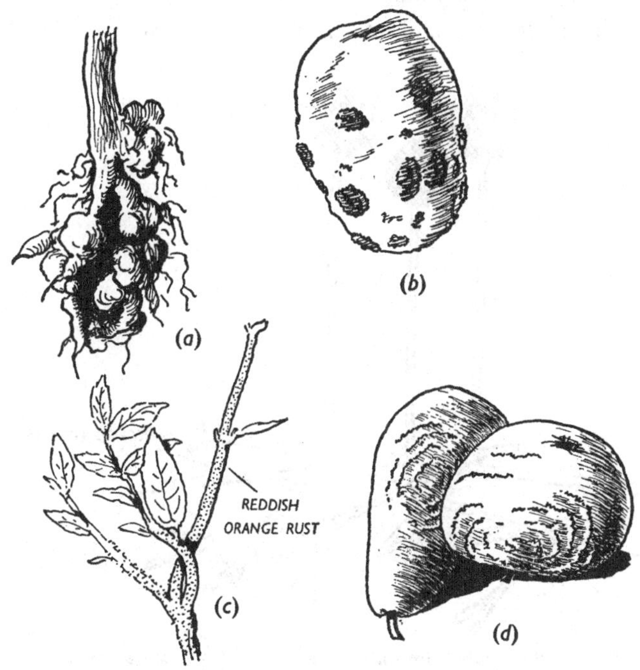

Fig. 8. *Parasitic fungi.* (*a*) *Clubroot in cabbage;* (*b*) *potato scab;* (*c*) *mint rust;* (*d*) *brown-rot in pears and apples*

Moulds

Another group of fungi that are very harmful are the moulds. Pin mould is a very familiar mould, which will grow on damp food that is covered with a bell jar. The food will gradually become covered with a white fluffy mould that is speckled with tiny black dots. Look at some of these white threads under the microscope and compare with Fig. 9. You will see that the hyphae branch in all directions and that they are not divided into cells as were the filaments of spirogyra. Short hyphae grow into the air and swell at their tips. Each swollen tip contains many nuclei and is cut off from the hypha by a cell wall. The tip becomes a SPORE BOX or

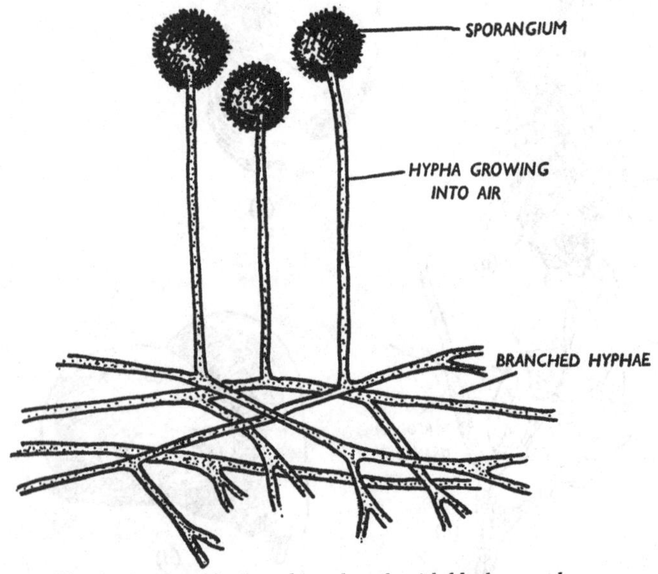

Fig. 9. *Pin mould Branching threads with black spore boxes*

SPORANGIUM inside which a number of SPORES are formed, each containing many nuclei. When the spores are ripe the sporangium bursts and the tiny spores are set free. They are scattered by the moving air and develop into hyphae when they alight on suitable damp food.

When the food is nearly used up, pin mould reproduces in another way. All the hyphae look the same, but there are two kinds which are called '+' and '−'. If a + hypha grows

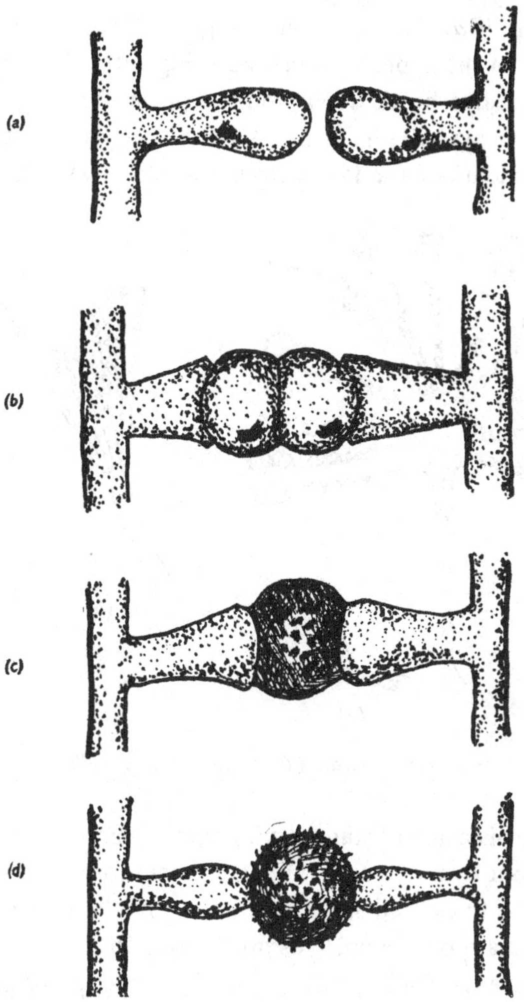

Fig. 10. *Pin mould: stages in the formation of a zygospore.* (*a*) *Lateral protuberances from long filaments;* (*b*) *tips of bulges cut off from hyphae;* (*c*) *fusion;* (*d*) *zygospore*

near to a – hypha, a bulge grows from each which touches the other (Fig. 10*a*). The tip of each bulge swells and is cut off from the rest of the hypha by a cell wall (Fig. 10*b*). The walls separating the tips break down and the nuclei fuse in pairs one from each hypha (Fig. 10*c*). A thick wall grows round the fused contents of the two bulges, forming a ZYGOSPORE

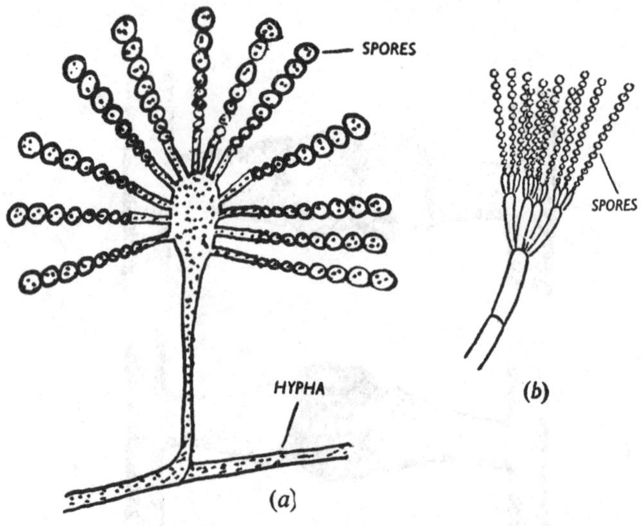

Fig. 11. *Blue-green moulds.* (*a*) *Eurotium;* (*b*) *Penicillium*

(Fig. 10*d*). This can withstand drying up or lack of food and is blown about by moving air. When it comes to rest on suitable food, the thick wall splits and a short hypha grows out which ends in a sporangium. This produces spores which have only one nucleus, which is either + or –. These spores grow into hyphae which are either + or –.

Blue-green moulds, also, are very commonly found on oranges and many other kinds of food, and some are even put into Gorgonzola and Stilton cheeses. Although these moulds are green they do not contain chlorophyll. If you have a microscope, look at some green mould under the high power and you will see the spores which grow from the ends of threads. There are two common types that you may see, Eurotium and Penicillium. Fig. 11 will enable you to recognize them.

Although moulds spoil our food, we have found, during recent years, that some moulds produce substances that will cure human diseases. Penicillin is obtained from one kind of Penicillium. Penicillin is used to cure several diseases, including septic wounds and abscesses.

Yeast

Yeast is a fungus that does not have hyphae, and it is very useful to us. Buy some yeast from your baker. You will think when you see it that it is not a plant. Put a very small amount

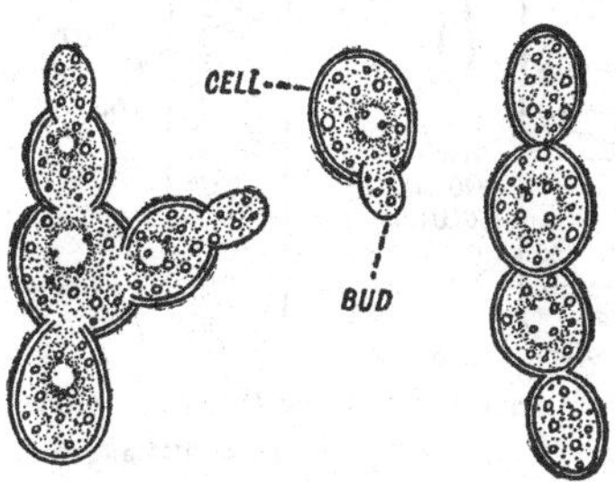

Fig. 12. *Yeast cells under the microscope. Successive stages of budding are seen, leading to the formation of chains of cells*

of yeast in some water on a slide and cover it with a coverslip. Then look at the yeast under a high power on your microscope. You will see that it consists of a lot of single cells (Fig. 12). When each cell is fully grown it reproduces by BUDDING. Small bumps appear on the outside of the cell, which later separate from the parent cell.

Yeast is very important because it FERMENTS sugar. If yeast is put into a sugary liquid it will use a very small amount of the sugar for food, and the remainder of the sugar it changes into alcohol and carbon dioxide. This change is called fermentation.

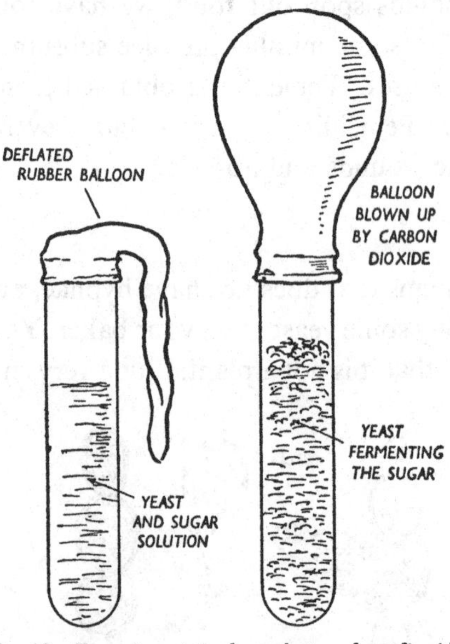

Fig. 13. *Experiment to show that carbon dioxide is given off when yeast ferments sugar*

Half fill a test-tube with sugar solution and add some yeast. Place a flattened balloon over the end of the test-tube, and leave for several hours. As the yeast ferments the sugar, a gas is given off which inflates the balloon (Fig. 13). This gas can be shown to be carbon dioxide as it turns lime water milky.

When bakers are making bread, they put yeast into the dough. This ferments the sugar in the flour and carbon dioxide is formed which makes holes in the dough and so makes the bread light. The small amount of alcohol that is formed disappears in the baking.

Yeast is used in brewing and in wine-making to ferment the sugar and so to produce alcohol. Carbon dioxide is also formed which can be seen as bubbles or as froth in wines and in beer. If alcoholic beverages do not contain carbon dioxide we say that they are flat. Brewers' and bakers' yeast is cultivated, but wine yeast appears as the 'bloom' on grapes. Dry wines are those in which all the sugar has been fermented, whereas sweet wines contain some sugar that has not been fermented.

Bacteria

Like the fungi, bacteria do not contain chlorophyll. We can only see them through a microscope, but we know that they are alive because they breathe, take in food, grow and reproduce like all living things. They also vary in behaviour when their environment changes (see chapter 9).

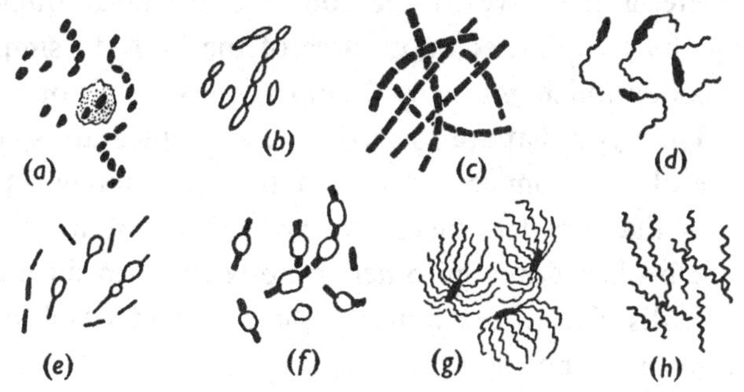

Fig. 14. *Various kinds of bacteria. Some are in chains. Spores are forming in* (*e*) *and* (*f*)

As they do not contain chlorophyll they cannot make their own food, so they have to live on living or dead animals and plants. For this reason many of them are harmful to mankind but a few are useful.

Bacteria are very, very tiny, about 1/25,000 inch long, and can only be seen through a powerful microscope. There are many kinds of bacteria which vary in shape and in size (Fig. 14). They consist of one cell only, which, like the amoeba, reproduces by splitting in two. If the bacteria are living in suitable surroundings they may split every twenty minutes. Some bacteria occasionally form thick walls around themselves and they are then called SPORES. Spores can resist heat or cold, and can be blown about from place to place. Other bacteria do not form spores, but they are carried about in the thin film of water that surrounds every particle of dust.

Useful bacteria

When animals and plants die they gradually rot away. Some bacteria that live in the soil feed on dead things, changing the complicated substances of the body to simple gases or solids which gradually disappear into the air or into the soil. The gases that are formed have an unpleasant smell which we always connect with dead things. Amongst the substances that are left there are compounds of nitrogen which are broken down by other bacteria to form nitrates, which will dissolve in the water in the soil, and which can then be absorbed by root hairs.

Look at the roots of a plant that belongs to the pea family. You will see little swellings or NODULES on the roots (Fig. 15). In these nodules live bacteria, which can use the nitrogen that is in the air and build it up into an organic form. Some of it passes into the plant, so giving it the nitrates that it requires. If a farmer wants more nitrate in his soil, he sows a crop of clover, as the roots have nodules containing these bacteria.

He eventually ploughs the clover plants into the soil and leaves them to rot. There are also bacteria living free in the soil which can change nitrogen into a soluble form.

Some bacteria help in the ripening of cheese and butter, in the tanning of leather, in the curing of tobacco, in the retting of flax, in the manufacture of alcoholic beverages and vinegar and in the disposal of sewage.

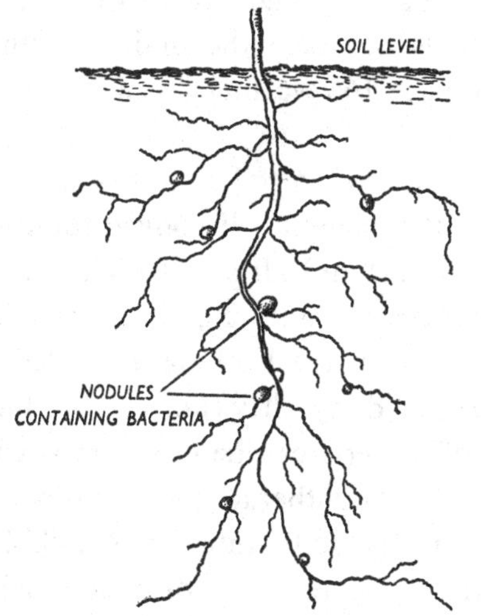

Fig. 15. *Nodules on roots of meadow vetchling*

Harmful bacteria

There are many bacteria, which we commonly call GERMS, that are very harmful because they get into our bodies, through the nose, throat or wounds, and cause diseases. Unless the skin is cut, no germs can enter the body through the skin, but occasionally germs get into a sweat gland and cause a pimple, or they enter a hair follicle and cause a boil.

We get rid of many germs when we cough or sneeze, or when we are sick, or if we have diarrhoea.

Germs affect our bodies in many ways. (1) Germs give out substances called TOXINS which poison our bodies. (2) The germs themselves are poisonous to us. (3) Germs may damage the cells of our bodies by going into them, or by making them go rotten. Nearly all germs spread throughout the body in the bloodstream, but the germs that cause diphtheria remain in the throat. These germs give out very strong toxins which pass into the bloodstream and are carried throughout the body. If these poisons reach the brain and the heart the patient dies. On page 166 we shall learn how the body kills the germs.

Bacteria may cause food to go bad or to putrefy. Food may be cooked to kill the bacteria. If cooked food is bottled or tinned whilst it is hot, and all air is excluded from it, the food will keep for a long time. Food will remain good in a refrigerator, not because the bacteria are killed, but merely because they are so cold that they cannot reproduce or rot the food. Colonies of bacteria can be grown in Petri-dishes (see Appendix B, p. 222). Diphtheria, tetanus or lock-jaw (whose germs are found in the soil), tuberculosis, cholera, plague, syphilis, gonorrhoea, septicaemia or blood poisoning and pneumonia are some diseases in human beings that are caused by bacteria.

Viruses

During recent years it has been found that there are some tiny things called viruses which are probably living and do not contain chlorophyll. They are so small that they cannot be seen even through a powerful microscope. With modern instruments scientists have been able to photograph them, to measure them, to separate them and to grow them. Like all living things they multiply. The study of viruses is very important, because they cause many diseases in plants and in animals.

Viruses in plants. It is known that viruses are spread from plant to plant in several ways. (1) By CONTACT. If a healthy plant that has a tiny wound in the leaf is touched by a plant that is infected with a virus disease it will become infected. The HEALTHY POTATO VIRUS and the TOBACCO MOSAIC VIRUS are spread in this way when one leaf is blown against another.

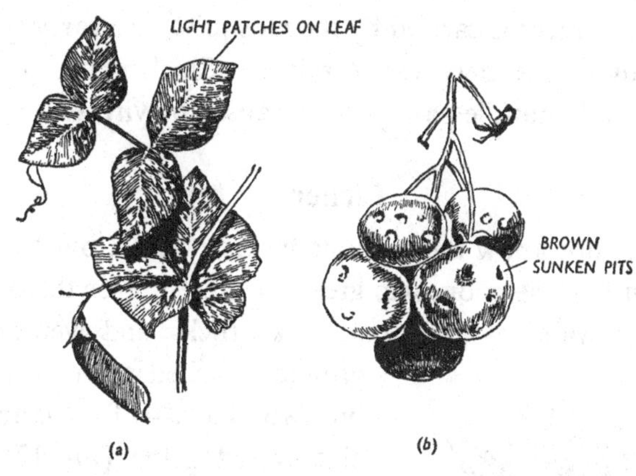

Fig. 16. *Plants attacked by virus diseases. (a) Pea mosaic, showing discoloration of leaves; (b) tomato streak, causing brown sunken pits on tomatoes*

(2) Many plant viruses are carried by insects. Aphides alone carry more than thirty virus diseases. They suck up the virus when they suck the sap of an infected plant, and then they inject it into the next plant on which they feed. Many potato and strawberry diseases are spread in this way, and the virus which causes the growth of tufts of twigs on many trees, called 'witches' brooms', is also carried by them. Leafhoppers also carry many diseases.

When a plant is attacked by a virus the cells may be killed or growth may be stopped, so that the plant becomes stunted. Sometimes growth is accelerated and growths are formed. The leaves of plants attacked by many viruses often lose their colour in patches (Fig. 16*a*).

Viruses in animals. Viruses cause many diseases in animals and in human beings. Foot and mouth disease of cattle, fowlpox and yellow fever are caught by contact, but smallpox, mumps, measles, influenza and colds are carried about in the air. Yellow fever is carried by the mosquito, lice carry typhus fever, and dogs may carry rabies. Chickenpox, german measles and infantile paralysis are caused by viruses.

Lichen

You may not know these plants by their name, but I expect that you have seen orange, green or greyish growths on old roofs, old walls, palings, tree trunks, rocks, and even on the ground. A lichen is not one plant, but two plants—a fungus and an alga living together (Fig. 17). The alga consists of single cells or of threads which are embedded in the hyphae of the fungus. As the alga contains chlorophyll it can make its own food, some of which is used by the fungus. In return for the food the fungus gives the alga water and minerals. Sometimes pieces of lichen become loose and grow into new plants, but the fungus and the alga may each form a kind of spore which only develop if they come into contact with one another.

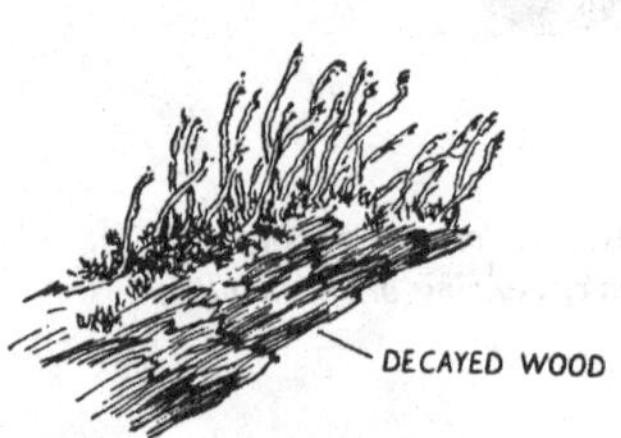

Fig. 17. *Lichen. Cladonia gracilis*

Mosses

Mosses are very simple little plants which grow in damp and wet places. They contain chlorophyll, and so can make their own food. Try to find different kinds of moss, and compare them with the one that is described here.

The commonest moss is the one that grows almost everywhere on waste land, footpaths, etc. Look carefully for this moss and try to find some that has little knobs or CAPSULES growing on long stalks (Fig. 18).

A moss plant has small hair-like roots (which are called RHIZOIDS), which fasten the plant to the soil. Tiny thin leaves grow in clusters round the short STEM. In the centre of the topmost rosettes of leaves grow tiny flask-shaped organs, which are too small to be seen without a microscope. In the flask-shaped organs of some shoots, tiny EGG CELLS are formed which eventually grow into the capsules or spore boxes. Before this happens, the egg cell has to be fertilized by a tiny SPERM which has formed in the flask-shaped organs of another shoot. These sperms correspond to the pollen grains in flowering plants which are carried by wind or by insects to other flowers in order that they may fertilize the ovules (see Book I). The sperms of the moss, however, are not carried by the wind or by insects; instead they swim by means of two little tails in the water after it has rained. The egg cells give out a chemical substance which attracts the sperms to them.

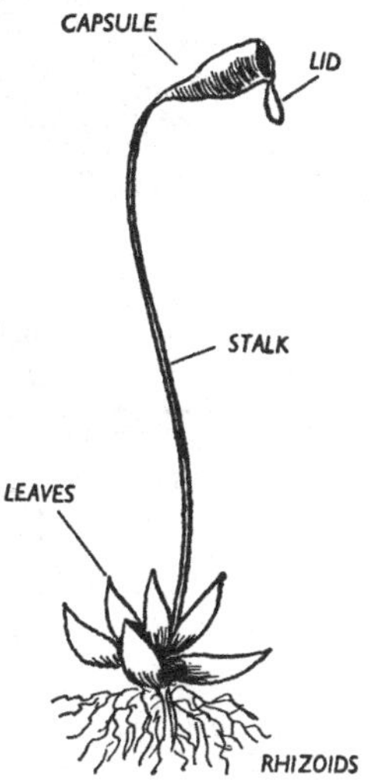

Fig. 18. *A moss plant*

From a fertilized egg the capsule or spore box grows. When the spores are ripe the lid of the capsule splits open and the spores are scattered. These spores eventually grow into new moss plants.

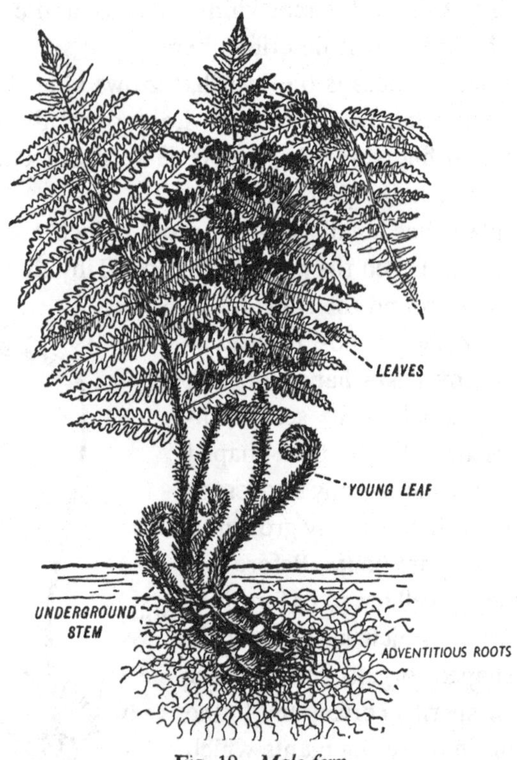

Fig. 19. *Male fern*

Ferns

Ferns are the largest of the flowerless plants. There are many varieties of fern that you could find during the autumn. Dig up a fern plant and look at it (Fig. 19). You will see that it has an UNDERGROUND STEM which is covered with brown

Fig. 20. *Hart's tongue fern*

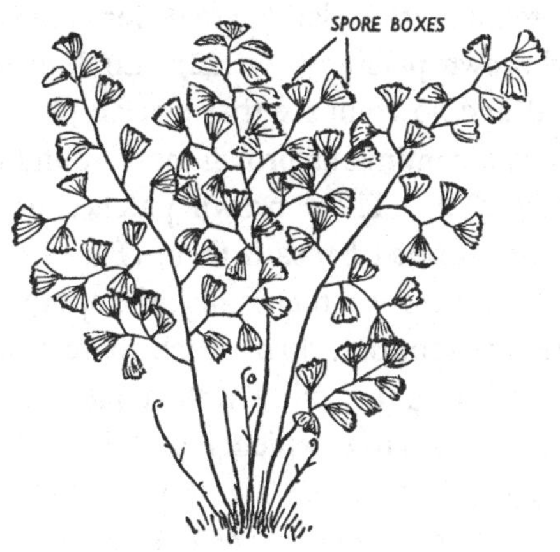

Fig. 21. *Maidenhair fern*

scales, and the remains of the stalks of old fern leaves. Many ADVENTITIOUS ROOTS grow from the underground stem. Each year, the old leaves die in the autumn and new ones grow again the following spring. Each leaf when it is young is covered with brown scales and it is curled up (Fig. 19). It uncurls as it grows. In the autumn look at the back of the

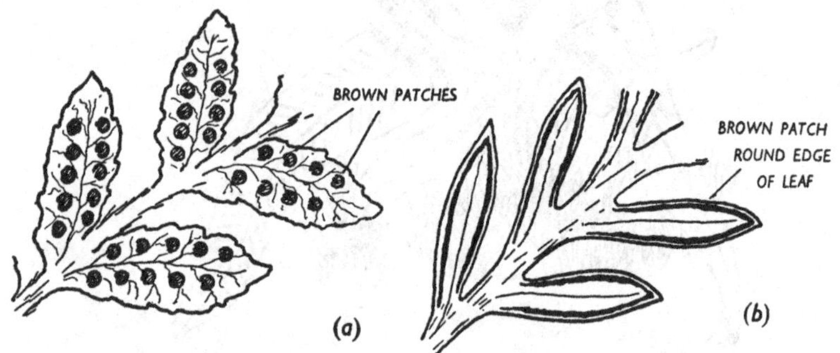

Fig. 22. *Parts of fern leaves showing the lower surfaces.* (*a*) *Male fern;* (*b*) *bracken*

leaf of the common fern called the male fern and you will see a number of brown patches (Fig. 22). Look at the back of a bracken leaf and you will see that instead of round brown patches there is a continuous brown mark round the edge of every leaf (Fig. 22*b*). These brown patches protect SPORE BOXES which are formed beneath them (Fig. 23). When the spores are ripe and the weather is dry, the spore boxes burst and the spores are shot out. From each spore a tiny leaf-like structure grows flat on the soil. On the underside of this tiny 'leaf' small rhizoids grow which fasten it to the soil. In addition there are flask-shaped organs similar to those found in the moss. Eggs are formed in the flask-shaped organs that are nearer to the middle of the 'leaf', and sperms are formed in those that grow round the edges. The sperms have many tiny hairs on them, which enable them to swim to the egg cells. When an egg cell is fertilized it grows into a new fern plant. Put some spores on to very damp soil in a jar covered with glass, and you will be able to see them develop into the tiny 'leaves'. Ferns contain chlorophyll and so can make their own food.

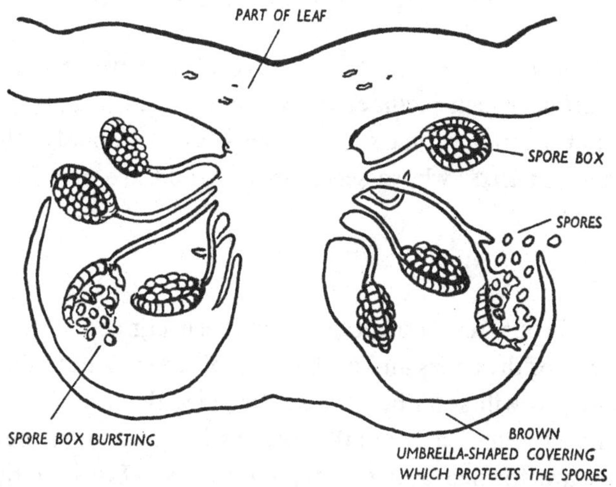

Fig. 23. *Section through a brown patch on a fern leaf*

CHAPTER 3

SEED-PRODUCING PLANTS

The seed-producing plants can be divided into two groups. First, a group which includes the CONIFERS, whose seeds are naked and are not protected by a seed box. Secondly, the FLOWERING PLANTS, whose seeds grow in and are protected by the seed box.

Conifers

All coniferous trees, except the larch, are evergreens. Make a collection of the cones and of the twigs of the trees described here, and you will soon be able to recognize them.

The SCOTS PINE is a very tall tree. As it grows taller, new branches are formed and the older ones die. Look at the leaves and you will see that they are needle-like and grow IN TWOS (Fig. 24*b*). Each pair of leaves grows from a dwarf twig. The leaves live for several years and, as new ones are formed each growing season, the tree is always green. These leaves are long and narrow and thick and leathery, and their pores, like those of the heather (see page 136) are sunk in little pits to prevent them from giving out much water. The CONES of all pine trees are made of WOODEN SCALES (Fig. 24*c*). Look at a cone that is open and you will see the seeds between these scales. Each seed has a brown papery wing attached to it, so that it can be scattered by the wind (Fig. 24*d*). Some people use these cones to forecast the weather, because they open when the weather is dry to enable the seeds to be blown away and they close when it is wet. Look at the tips of the branches in May and you will see clusters of buds. Some of these clusters are yellow in colour and consist of a number of

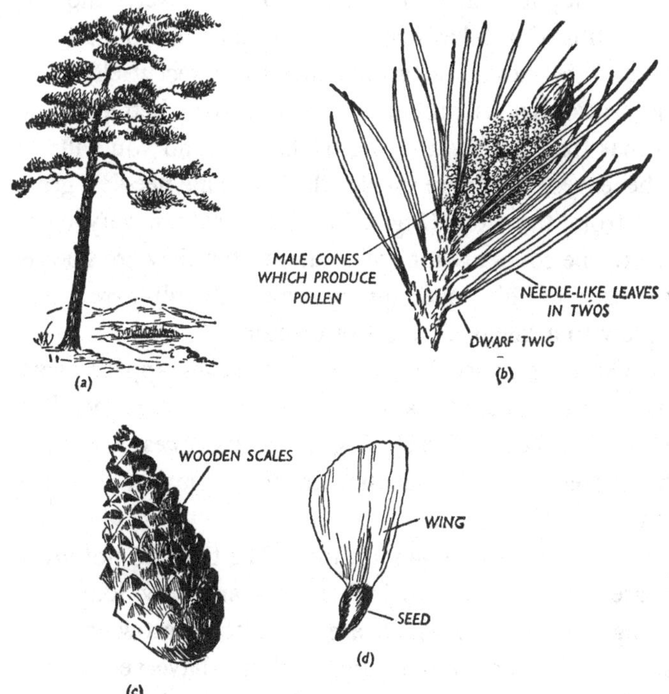

Fig. 24. *Scots pine.* (*a*) *Tree;* (*b*) *shoot with male cones;* (*c*) *old female cone;* (*d*) *a single seed*

scales that overlap one another (Fig. 24*b*). Pull one to pieces and look at it through a lens. You will see a little bag on the lower side of the scale. This is really an ANTHER filled with POLLEN. Each pollen grain has two little air bladders attached to it to make it float in the air. On the same tree you will see bigger female cones which grow in twos or threes. Look at one and you will see two OVULES on the upper surface of each scale. On a dry day, when the pollen is being blown about by the wind, the scales open to enable the pollen to reach the ovules. The pollen grains do not fertilize the ovules immediately, but they remain in the female cone for twelve months before fertilization takes place. During this time the cone is closed and secretes a sticky substance to protect itself.

The SPRUCE FIR is probably known to you better as the Christmas tree. Look at a twig of this tree and you will see that the leaves, which are shorter than the pine needles, grow SINGLY from the stem (Fig. 25*a*), but they are very close together. The cones are long and narrow, but they are SOFT as they consist of thin, overlapping scales. A full-grown tree may grow to a height of 160 feet or more.

The LARCH is a deciduous tree. Its leaves appear every spring as bright green tufts of 15 to 20 needles, arranged like the hairs on a sweep's brush (Fig. 25*b*). Its cones, like those of the fir tree, are soft, but they are much smaller than the fir cones.

CEDARS are not very common trees. The trees do not grow to a great height, but their branches are very long and sweeping. The leaves, which are evergreens, grow in little tufts. The cones are smooth and oval and have very broad scales. When the seeds are ripe the cone falls to pieces (Fig. 25*c*).

CYPRESS trees are used to ornament many gardens. There are many varieties which vary in colour and in the shape of the leaves. The leaves are flattened and lie close to the stem. The cones are very tiny. The male cones have an enormous quantity of pollen (Fig. 25*d*).

YEWS are often grown in churchyards. They live to a great age and grow to an enormous size. The leaves, which are

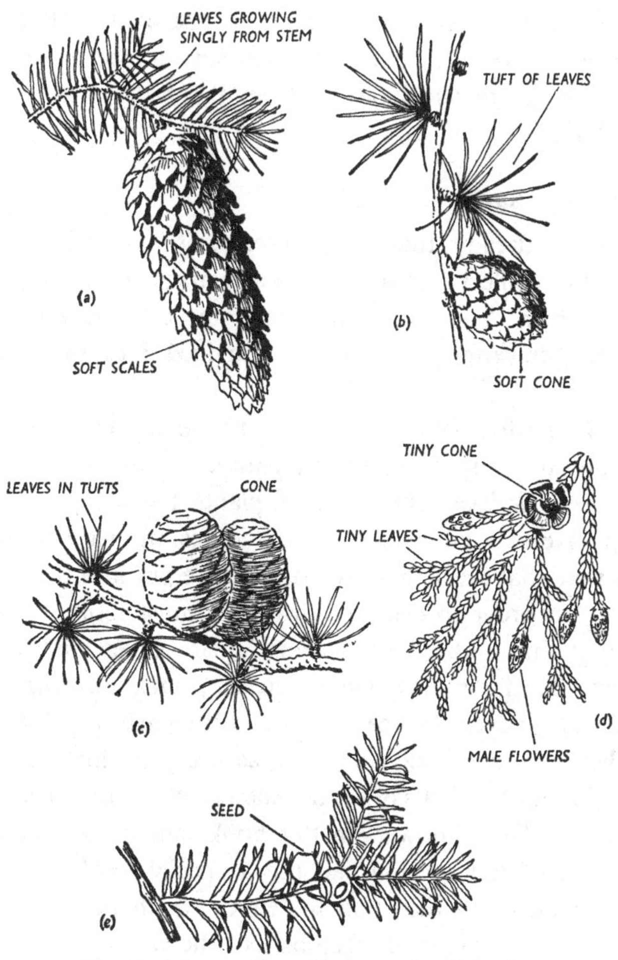

Fig. 25. *More conifers. (a) Spruce fir; (b) larch; (c) cedar; (d) cypress; (e) yew*

narrow and short, grow round the stem in spirals. The pollen and the ovules are produced by different trees during February and March. Look at a male flower and you will see that it has ten stamens. The pollen is blown by the wind to the female flowers, each of which contains one ovule. No cone is formed. The single seed becomes surrounded by a red fleshy outgrowth which grows from the base of the seed (Fig. 25*e*).

Flowering plants

We have already studied the structure of a flowering plant in Book I. Collect as many flowers as you can and look carefully at them. You will notice that some of them are very similar. Flowering plants can be divided into two main groups:

(1) *Monocotyledons*. The seeds of these plants contain only one cotyledon and the leaves are parallel-veined.

(2) *Dicotyledons*. The seeds of plants belonging to this group have two cotyledons and the leaves are net-veined.

Flowers that are similar in structure are put into the same family. In order to classify flowers we must look at them carefully and notice their stems, their leaves and their flowers. Find out how many sepals, petals, stamens and pistils they have. Notice whether these parts are free or whether they are joined together as are the sepals of the primrose and the petals of the deadnettle. We cannot mention all the families of flowers in this book, but if you make a collection of flowers you can use a special book called a FLORA, which will help you to tell to which families they belong. For instance, all members of the wallflower family can be recognized because they have four sepals, four petals, and four long and two short stamens. Members of the deadnettle family have square stems and flowers that are similar in structure to that of the deadnettle.

CHAPTER 4

ANIMALS: INVERTEBRATES

We have studied many animals that we have found in fresh water, in our gardens and fields, and in our countryside. We must now learn something about the families to which they belong, and also learn a little about a few animals that we have not yet studied, but which belong to these families.

We have already learned that animals can be divided into two big groups: (1) the INVERTEBRATES, which have no backbone; (2) the VERTEBRATES, which have a backbone. In this chapter we shall study the invertebrates.

Single-celled animals or protozoa

The smallest and simplest animals consist of only one cell, which is able to feed, breathe, get rid of waste material, move and reproduce. There are many animals that belong to this group. Most of them, like the amoeba, live in fresh water and are usually free-living, but they may live in colonies. If you can find some water weeds that seem to be covered with a whitish film, look at a small piece of it under a microscope. You will see that it consists of many single-celled animals, each of which is on the end of a long stalk, which contracts by coiling up like a spring when it is touched (Fig. 26*a*). It is called a 'bell' animal or VORTICELLA. The small hair-like things round the mouth, which are called CILIA, drive particles of organic matter into the mouth. Vorticella digests its food in the same manner as the amoeba. When full grown the bell splits in two, lengthwise. One half remains attached to the stalk and grows, whilst the other half grows some cilia at

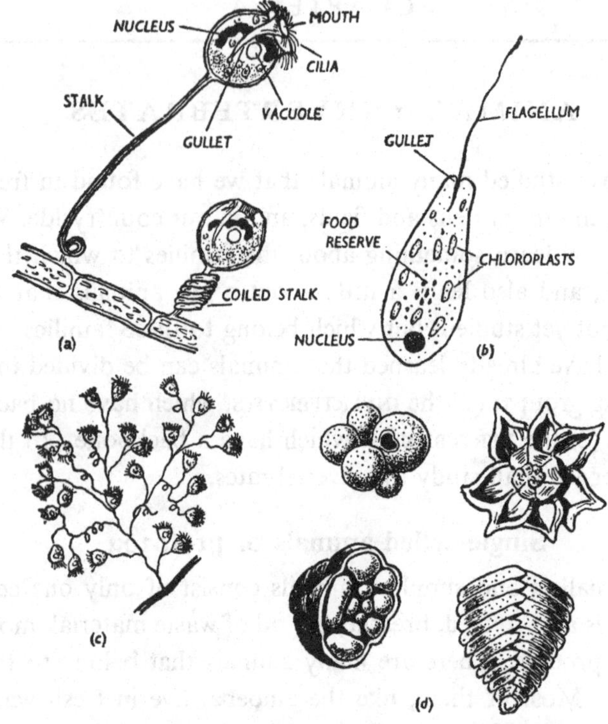

Fig. 26. *Single-celled animals.* (*a*) *Vorticella;* (*b*) *euglena;* (*c*) *colony of carchesia;* (*d*) *shells of chalk*

one end, swims away and finally settles down and grows a new stalk. In CARCHESIUM (Fig. 26*c*) the stalk also splits, and the daughter remains attached to the parent. So a colony is formed.

If you put a spot of pond-water on a slide (see Appendix B, page 219) and look at it through a microscope, you will be surprised at the number of single-celled animals and plants that you will see floating in the water (Fig. 26*b*).

Many single-celled animals that live in the sea secrete a shell round themselves (Fig. 26*d*). They float freely in the sea when they are alive, but sink slowly to the bottom of the sea when they die. These chalky cells have accumulated for centuries and have formed a thick layer which has gradually consolidated to hard chalk rock under the pressure of the ocean. We know that the crust of the earth moves, and that sometimes land which was once below the sea comes above the sea level. When this happens, as it has along the chalk cliffs of South-east England, we are able to see these rocks.

A few single-celled animals live inside the bodies of animals. They are usually carried there by other animals. Sleeping sickness (which is carried by the tsetse fly) and malaria (which is carried by the mosquito) are both caused by amoeba-like animals.

The hydra family or coelenterates

The simplest of these animals consist of many cells which are arranged in two layers enclosing a sac-like cavity, the gut cavity or ENTERON, which has only one opening—the MOUTH. The hydra, if you remember, is like this. If you look carefully in rock pools along the sea shore you will find ANEMONES, which look like very big hydras. When they are not covered by the sea water they look like lumps of brown, blue, red or green jelly which are fastened to the rocks. Look at them again when they are under water and you will see that they have a ring of TENTACLES round the mouth, with which they catch their food. Many tropical anemones are big enough to catch small fish. If you can get some sea water, set up an aquarium as shown in Book I, using sea weeds instead of fresh-water weeds. Place sand at the bottom and put stones on top of the sand, as many sea animals attach themselves to stones (Fig. 27).

Hydra and anemones are similar in many ways, but they reproduce in different ways. The sperms and the eggs are formed in the same animal in hydra, and they are said to be

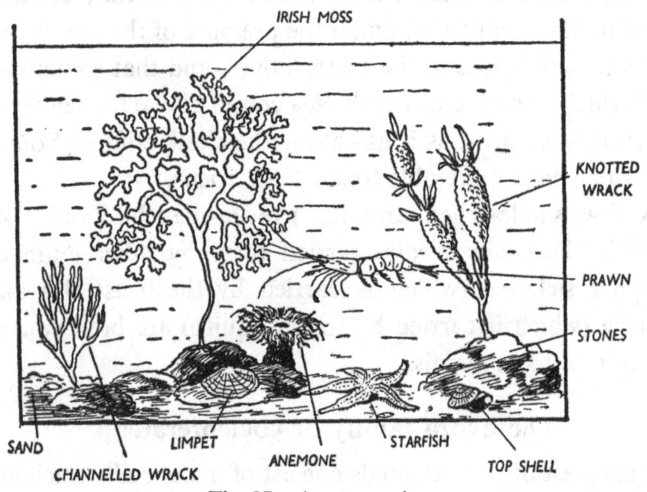

Fig. 27. *A sea aquarium*

HERMAPHRODITE. Anemones are UNISEXUAL, because the eggs and the sperms are formed in different animals. The eggs in the anemone are formed inside the animal, and not on the outside as they are in the hydra. When ripe, the sperms swim about in the water and are attracted by some chemical substance to the female. They swim in through the mouth into the GULLET, where the eggs are developing in the ovary. The young sea anemones remain inside the mother's body for a time until they have tiny tentacles. Then they swim out through the mouth and finally attach themselves to a rock.

Some anemones fasten themselves on the shell of a hermit crab (see page 66). The crab carries the anemone about; in return the anemone protects the crab by partly hiding it (Fig. 28).

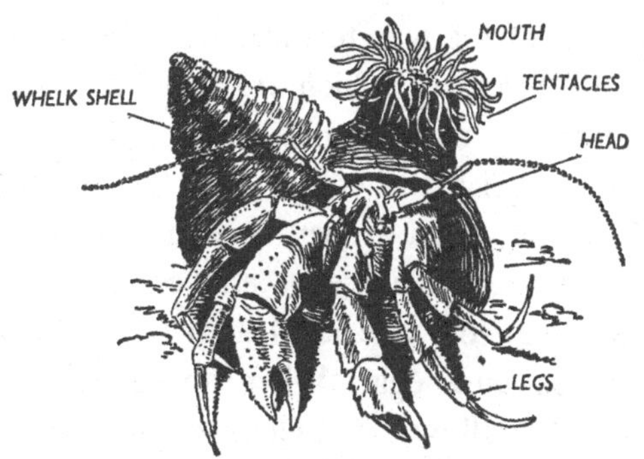

Fig. 28. *Anemone on empty whelk shell, inside which a hermit crab is living*

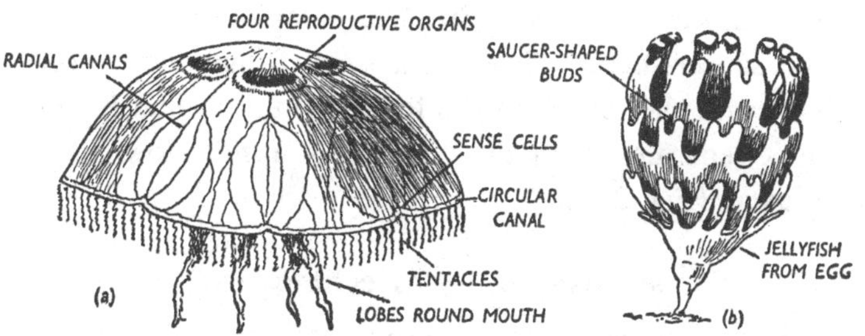

Fig. 29. *Jellyfish.* (*a*) *Side view;* (*b*) *young jellyfish budding*

In some places around the coast (for instance, in North Wales) you can see saucer-shaped pieces of transparent jelly moving about in the water. These are JELLYFISH, which vary in size from 2 to 12 inches. Look at a jellyfish and compare it with Fig. 29. The edge of the jellyfish is fringed with TENTACLES. This fringe is broken in eight places, where there are special SENSE CELLS with which the animal is able to tell if there is anything that is worth eating near to it. You will see the MOUTH on the underside. It is surrounded by four LOBES on which there are stinging cells that are similar to those on the tentacles of the hydra. The body cavity only stretches for part of the way towards the edge of the body; it then sends out branches, which are called RADIAL CANALS, to the end. Here they all join up to form a CIRCULAR CANAL.

Look at your jellyfish and you will see four whitish horse-shoe-shaped organs which are easily seen through the jelly. These are the reproductive organs and contain either EGGS or SPERMS. Sperms from one jellyfish swim in the water to a jellyfish that contains eggs. When an egg is fertilized it grows into a little hollow sac, then it swims away from its mother. It may develop into a jellyfish or it may fasten itself to a rock and form BUDS. These buds do not form at the side of the sac as they do in the hydra, but on top of the sac. These buds are saucer-shaped. Several buds may form on top of one another, with the oldest on the top. When each bud is full-grown it separates from the rest, turns upside down, and swims away (Fig. 29*b*).

The starfish family or echinoderms

The animals that belong to this group have not only a cavity in the body, but also a separate digestive system along which the food passes. If you look on the sea shore you may find many five-rayed STARFISH which have been washed ashore.

Look at a starfish and you will see that it has a CENTRAL DISC from which the five ARMS or RAYS grow (Fig. 30). The upper side of the starfish is very rough, as it consists of hard plates from which tiny SPINES project. Between these spines there are tiny PINCERS, which are continually snapping to keep anything away from the starfish that might be harmful. On the upper side of the central disc you will see a small round mark, where water enters the body. There is a special WATER SYSTEM. The water passes through a tube into a RADIAL CANAL which goes round the edge of the disc. From this radial canal a canal passes into each arm.

Look at the underside of a starfish and you will see a groove going along the middle of each ray. On each side of these rays there are two rows of TUBE FEET which are connected with the

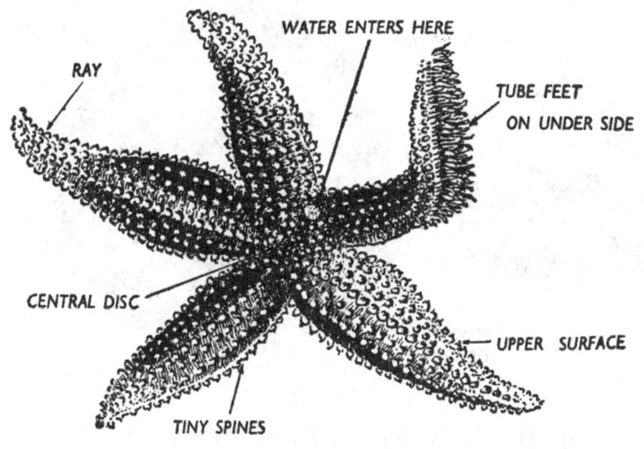

Fig. 30. *Starfish*

water canal. The feet extend when water enters them, or they can be withdrawn when the water passes out of the tube feet. Each tube foot ends in a SUCKER. It is with these feet that the starfish moves about, and it can move in any direction. The feet on one ray cling to a rock, whilst the remainder of the body is dragged up to this ray. The MOUTH is on the underside and it leads into a wide STOMACH, which can be pushed out through the mouth to pick up food. Starfish eat animals such as mussels, which have a double hinged shell. A starfish walks on top of a mussel and pulls the two parts of the shell apart with its suckers. It can then pick up the soft body of the mussel with its mouth. Food passes from the stomach through tubes into the rays. Waste food is not got rid of through the mouth; instead it passes out through a small opening called the ANUS, which you will see on the top of the disc.

The SPERMS and the EGGS are shed by separate animals into the water, where fertilization takes place. There are five pairs of OVARIES (where the eggs are formed) or TESTES (where the

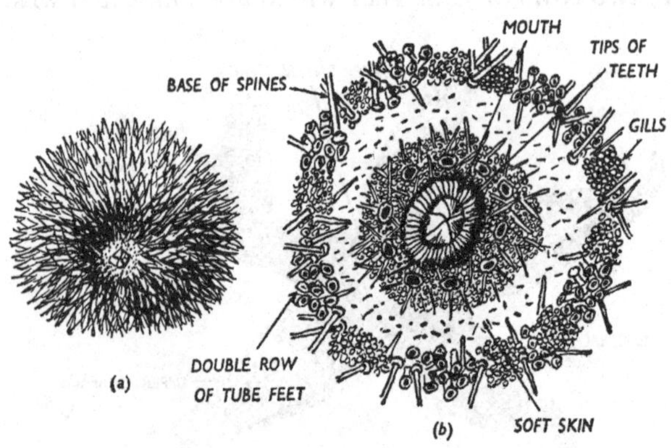

Fig. 31. *Sea urchin. (a) View from underside; (b) mouth region enlarged*

sperms are formed) in the disc part of the body. If you cut a starfish into large pieces each piece will develop into a complete animal.

Starfish have special small, thin-walled outgrowths from the skin through which oxygen can pass into the body cavity.

SEA URCHINS look like pale, greenish-purple, prickly balls (Fig. 31). The SHELL, which is hardened with lime, is covered with long, movable SPINES. Between these spines there are several kinds of PINCER which help to keep the shell clean. If you can get a living specimen, put it in a tank of sea water. Gently pass a fine thread over it and you will see these pincers snapping. There are five double rows of TUBE FEET, similar to those of the starfish, which can be pushed out a long way. It is on these that the sea urchin walks. It has five enormous TEETH, the tips of which protrude slightly through the MOUTH. The skin around the mouth is not hardened, so the teeth can be pushed farther out of the mouth to tear lumps off the sea-weed or the small animals that they eat. At the outer edge of this softer part there are five pairs of branched GILLS through which the sea urchin breathes. The EGGS and the SPERMS are produced by different animals. They are shed into the water where fertilization takes place.

Worms

There are three large groups of worms which differ very much from one another: (1) FLAT-worms; (2) ROUND-worms; (3) SEGMENTED worms.

Flat-worms

Some flat-worms live in fresh water. They are rather small and are not easily seen in a pond unless the water is clear and still. Bring some weed and water back from a pond and put it into your fresh-water aquarium. You may later see some black or grey flat-worms creeping up the side of the tank. Look at one with a lens and you will see that its body is flattened from top to bottom. It has two EYES at the blunt end of the body. The MOUTH is in the centre of the lower side of the body (Fig. 32). It leads into a STOMACH, from which one tube goes to the front end of the body and two tubes go

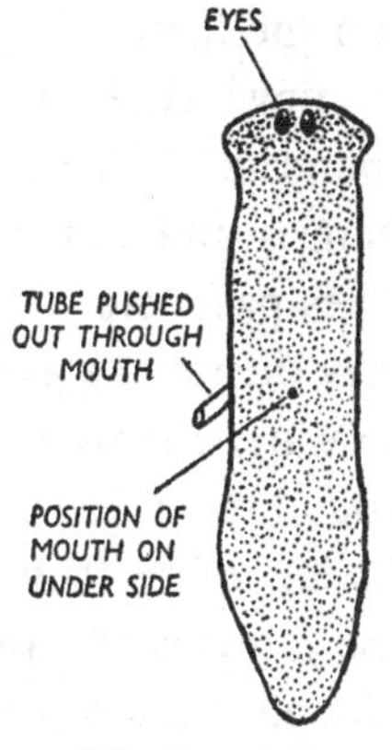

Fig. 32. *A flat-worm. Planaria*

to the hinder end. These tubes branch repeatedly and carry food to all parts of the body. Waste food has to pass back along these tubes and pass out through the mouth. Flat-worms eat small animals or dead bodies, which they tear to pieces with a tube that they can push out through the mouth. Watch a flat-worm moving along the glass of the aquarium. It seems to glide along. This gliding movement is brought about by tiny hair-like things called CILIA, which cover the body and which are continually moving.

Nearly all flat-worms have both male and female organs in the same animal. Each individual when mature lays about ten eggs at a time in a tiny capsule which contains food. Each capsule is fastened to some object in the water.

LIVER FLUKES are similar to these flat-worms. They live in the liver of sheep causing great damage and sometimes death. The eggs of the liver fluke pass from the liver of the sheep into the bile duct, thence through the alimentary canal and out with the faeces. If they drop in a damp place they develop into tiny larvae which can swim. At this stage they must enter a pond snail. In the snail's body each larva undergoes several changes and multiplies rapidly. Finally many young liver flukes pass into the water and swim about for a short time. They then make a case round themselves and may remain in the water for twelve months, or they may remain on plants around the edge of a pond for a few weeks. They will eventually die unless they are eaten by sheep. Inside the sheep, the case is digested, and each young liver fluke bores its way through the sheep's body until it reaches the liver.

The TAPEWORM, which lives in the human food canal, is also a flat-worm. It looks like a long piece of white ribbon which tapers to a point at one end, the HEAD END. On the head there is a row of hooks, and there are four SUCKERS with

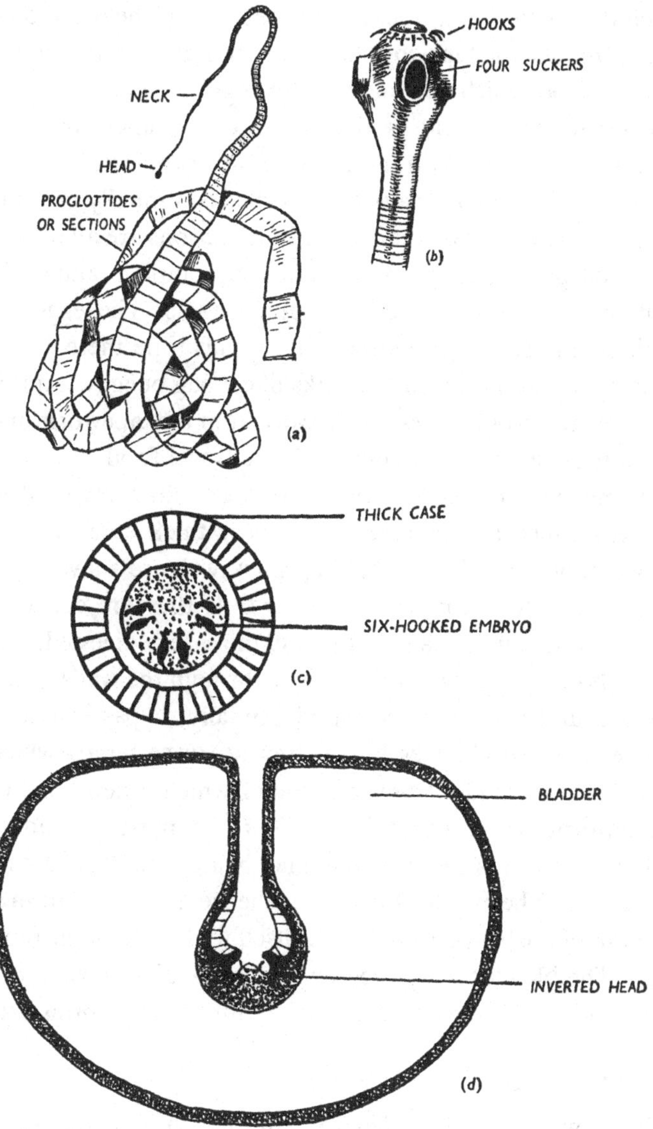

Fig. 33. *Tapeworm.* (*a*) *Full grown tapeworm;* (*b*) *head;* (*c*) *six-hooked embryo;* (*d*) *section through bladder-worm*

which the tapeworm fastens itself to the side of the food canal (Fig. 33*b*). As the tapeworm is so thin it is able to absorb the digested food which surrounds it through its skin.

A tapeworm's body is divided into SECTIONS (or PROGLOTTIDES). There may be as many as 1000 proglottides in all. Behind the head there is a NECK, which is continually growing and giving rise to new sections. All the sections are alike and, when full grown, each one is almost a complete animal in itself, as it contains both EGGS and SPERMS. The eggs are fertilized inside the proglottis, and by this time there are so many eggs that the proglottis looks like a bag of eggs. At this stage several proglottides break away from the tapeworm and pass out of the person's body with the waste food.

Meanwhile the eggs have developed into six-hooked embryos which are covered with a very thick case. Everything except the embryos and their thick walls rots away, but these may survive for several months. If they are eaten by the right host, such as a pig or cattle, the case is digested, and the six-hooked embryo bores its way through the walls of the stomach and goes into the blood. Finally, it passes into a muscle where it changes into a tiny BLADDER-WORM which consists of a bladder from one side of which a new head of a tapeworm grows inverted (Fig. 33*d*). If meat containing a bladder-worm is eaten by a human being, and the bladder-worm has not been killed by cooking the meat, the head pushes itself out of the bladder and fastens itself to the wall of the food canal. The bladder is digested and the neck gives rise to new proglottides. You now realize how harmful these worms are.

Round-worms

These worms vary in length, but they have long, thin, round bodies that are not divided into segments. A few round-worms live freely in fresh water or in the soil, but nearly all of them are parasites living entirely in the bodies of other animals or plants. One very common round-worm is the THREADWORM that lives in the lower end of the food canal of children or even adults. Dogs, too, frequently have these threadworms. ASCARIS is a much larger round-worm, which may be 16 inches long. One kind lives in the food canal of man, and another kind lives in the stomach of the horse. All these worms are very harmful (Fig. 34).

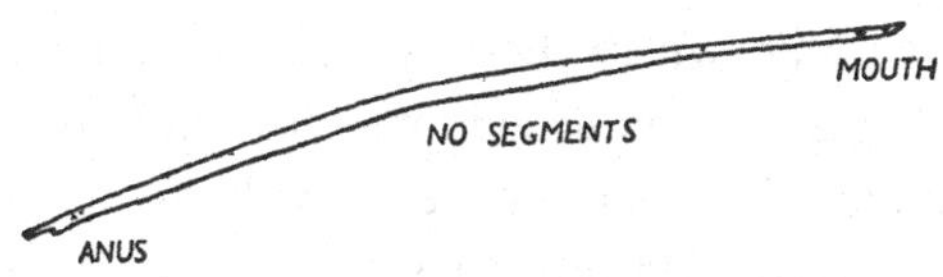

Fig. 34. *Human ascaris*

Segmented worms

The worms belonging to this group have long, thin, segmented bodies which are more complicated in structure than the other worms. They are chiefly free-living. As we have studied the earthworm and the horse leech, which belong to this group, we need not say more about them here.

The snail family or molluscs

We have already studied several animals that belong to this group and we know that, with the exception of the slug, all these animals can be recognized by their shells. The shell is formed from secretions of a fold of skin, called the MANTLE, which covers the body. The mantle adheres closely to the shell and, round the open edge of the shell, it forms a thick rim called the COLLAR. The shell consists of three layers: (1) the outer, horny layer; (2) the thick middle layer which consists of a substance called calcium carbonate, (3) the inner, pearly layer of mother-of-pearl which is a very thin layer of calcium carbonate that is arranged in layers. The inner, pearly layer can be secreted by any part of the mantle, but the other two layers are only secreted by the collar. As the animal grows, the shell is added to by the collar. The substances added at each growth period may vary a little, so lines appear on the shell.

Place a shell in 10 per cent hydrochloric acid. You will see that it dissolves, leaving only the outer horny layer. If snails live in stagnant water, where there is much decaying matter, the water will contain carbonic acid which would dissolve the calcium carbonate if it were not covered with the horny layer.

Molluscs can be divided into two groups: (1) the SINGLE-shelled animals or UNIVALVES, such as the snail, whose shells are in one piece as they are formed from the mantle which covers the body in one piece; (2) DOUBLE-shelled animals or BIVALVES, such as the swan mussel (see Book I), whose shells consist of two pieces which are hinged together. In these animals the mantle hangs down on the two sides of the body as separate lobes, each lobe giving rise to one piece of the shell. In this book we shall learn a little about the molluscs that live in the sea.

Single-shelled animals or univalves

All these animals are similar to the pond snail in structure, and they eat seaweed with their rough tongues. Each one has a well-developed head with two eyes and two feelers, and a foot on which it glides from place to place. The FOOT can also be used to fasten the animal to the rocks to prevent it from being washed away by the sea. No sea molluscs come to the surface of the water to breathe; instead they breathe through a special GILL which lies in the space between the body and the shell. Some of these animals have a HORNY PLATE on the upper surface of the foot, which completely closes the opening of the shell when the animal has withdrawn into it. (The

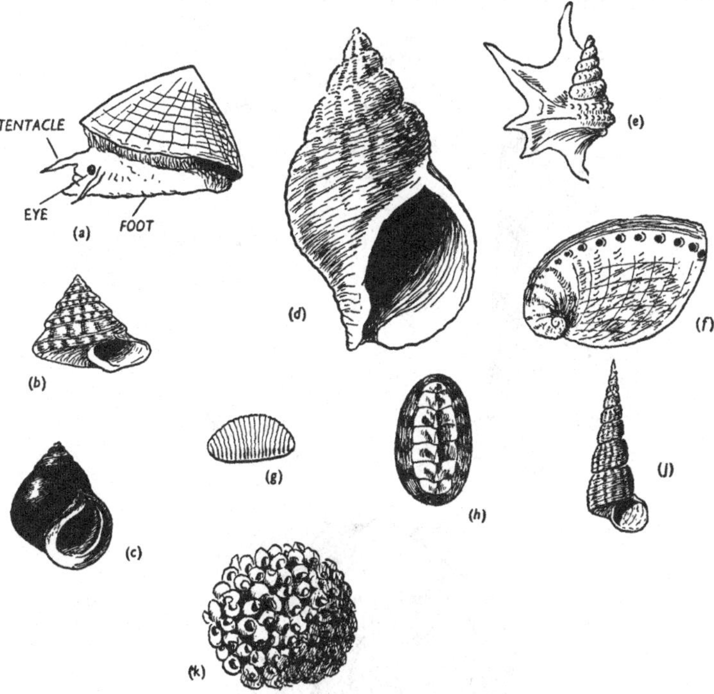

Fig. 35. *Univalves found in the sea.* (*a*) *Limpet;* (*b*) *top shell;* (*c*) *periwinkle;* (*d*) *wide-mouthed whelk;* (*e*) *pelican's foot;* (*f*) *ormer;* (*g*) *cowrie;* (*h*) *coat-of-mail;* (*j*) *auger;* (*k*) *whelk's egg cases*

VIVIPAROUS POND SNAIL, which lives in fresh water, also breathes through a gill and has a horny plate. These snails do not lay eggs but produce about fifty fully developed young at a time.) The shells of WINKLES, WHELKS, LIMPETS and COWRIES are often picked up on the sea shore. Fig. 35 shows you the shells of several single-shelled animals that you may find on the sea shore. See how many of them you can find. If you have a sea-water aquarium, get some living specimens out of rock pools and watch them carefully. You will probably find the empty egg cases of whelk eggs on the sea shore (Fig. 35*k*).

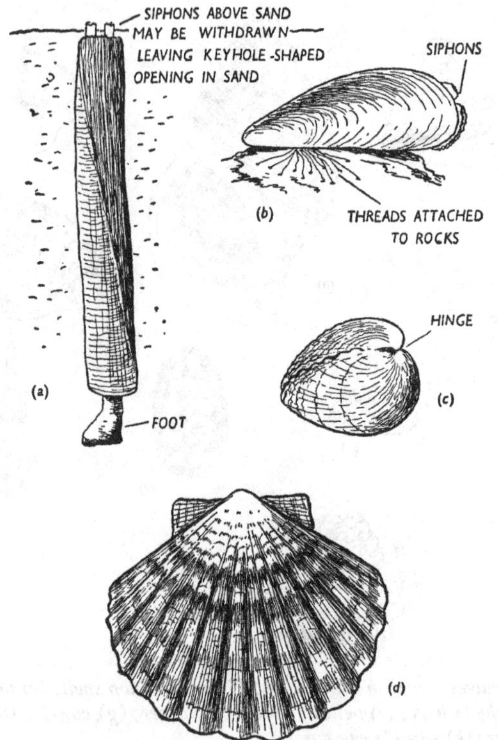

Fig. 36. *Bivalves found in the sea. (a) Razor shell; (b) mussel; (c) cockle; (d) scallop*

Double-shelled animals or bivalves

Many bivalve molluscs live in the sea. Look for the shells of these animals along the sea shore, and try to find the two halves of the shell hinged together. When the animals are alive it is difficult to open the shells, as they are held together by strong muscles, but they soon open when the animal is dead. The commonest bivalve mollusc to be found is the bluish SEA MUSSEL, which is fastened to a rock by means of many THREADS which grow out of the hinge (Fig. 36*b*). These threads, which prevent the mussel from being washed away, can be unfastened from the rock and the mussel can walk from place to place on its FOOT, which is similar to that of the swan mussel. Halves of COCKLE shells are very plentiful on many beaches, but you seldom find a live cockle because it lives buried in the mud or the sand between high and low tide marks (Fig. 36*c*). All bivalve molluscs have SIPHONS for breathing and for feeding similar to those of the swan mussel. The cockle pushes its two long siphons out of the sand into the water. It moves about on its FOOT. An OYSTER cannot move, as the left half of its shell is cemented to a rock. This half is like a basin, whilst the top half is flat. There are special oyster breeding beds in Whitstable. The RAZOR shell burrows vertically into the sand, pushing its two siphons up into the water (Fig. 36*a*). SCALLOPS have a lower saucer-shaped shell and an upper flat shell (Fig. 36*d*).

Arthropods

The crab family (CRUSTACEA), the spider family (ARACHNIDS), millipedes and centipedes (MYRIAPODS), and INSECTS all belong to this large group, as they have many things in common. They have JOINTED OR SEGMENTED BODIES and JOINTED LEGS. They MOULT, as their bodies are covered with a substance that cannot grow. Their bodies are usually divided into three parts: (1) HEAD; (2) THORAX; (3) ABDOMEN.

Crustacea

Nearly all the members of this family have a body that is covered with a hard substance called CHITIN, which not only protects the body but also supports it and acts as an outer- or exo-skeleton. Although the body is segmented, the exoskeleton of some segments may fuse together dorsally. Look at a crab, a prawn, a lobster or a shrimp and you will see that this is so (Fig. 37). Each segment has a pair of leg-like things called APPENDAGES, which may be used for catching or eating food, for walking or for swimming. They usually have two eyes which are on the end of long stalks. Tiny crustaceans breathe through their skin, but the larger ones have GILLS, which are attached to either the front or the back appendages.

We have already studied fresh-water shrimps, cyclops, water fleas, water slaters and wood lice. Many Crustacea live in the sea. You can easily find small crabs in rock pools along the sea shore, or you can catch shrimps in the shallow water if you use a special shrimping net.

SHRIMPS, PRAWNS, CRAYFISH (which live in streams) and LOBSTERS are very similar in structure. Try to get specimens, alive if possible, of these animals and look at them carefully. When alive they vary in colour from a pale greenish-grey to brown. Prawns are transparent when swimming about, but they change colour to match the colour of the seaweed on which they are resting. These animals only become bright pink in colour when they are boiled. The hard covering on the segments of the HEAD and of the THORAX is in one piece on the dorsal side, and forms the CARAPACE (Fig. 37*a*). The EYES are stalked and the FEELERS are very long. There are six pairs of peculiarly shaped appendages or JAWS round the mouth which enable the animal to catch and to eat its food.

There are five pairs of WALKING LEGS on the thorax. The first pair usually have large PINCERS; in prawns and in shrimps the pincers are on the second pair. On the abdomen there are

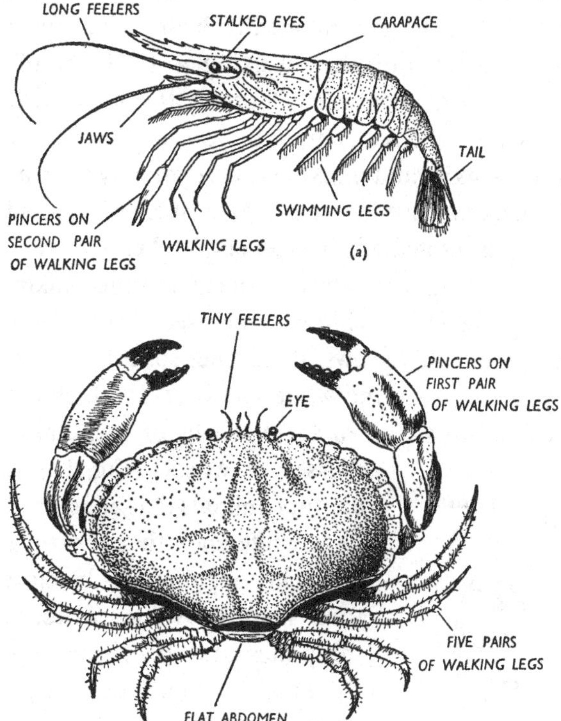

Fig. 37. *Crustacea. (a) Prawn; (b) crab*

five pairs of SWIMMING LEGS and a TAIL which help the animal to swim. The abdomen is usually bent (Fig. 37*a*). When swimming the animal straightens and then bends its tail to push itself backwards through the water. The female carries the fertilized EGGS on her swimming legs until they hatch. If you buy dead shrimps or prawns from a fishmonger you may find some of them with eggs attached to their legs. Look at a CRAB and you will see that it is very different from the other Crustacea. The THORAX is very wide, and the ABDOMEN is short and thin, and is permanently bent under the body. Crabs walk sideways on their walking legs. If one of these animals loses a leg, a new one will grow. It is not easy to find HERMIT CRABS on the sea shore, but you may find dead ones amongst the winkles and the whelks that you buy from a shop. The abdomen of a hermit crab is soft, as it is not covered with chitin, so it pushes its body into an empty shell that is just the right size for it (Fig. 28). Only the head and the walking legs (which are protected by chitin) can be seen protruding from the shell. As its body grows, it searches for a bigger shell.

Have you ever seen the white BARNACLES, which are permanently fixed to stones or to the shells of such animals as mussels or whelks (Fig. 1*b*)? These are tiny animals that are surrounded by a white shell that is open at the top. This opening can be closed by four small movable plates. When the plates are open, small feathery appendages come out and wave about in the water catching any food that they touch. This movement of the appendages helps to bring more oxygen to the animal. Have you ever found tiny flat animals which seem to be fastened to a stickleback? These are CARP-LICE, which are about one-eighth of an inch long. They have two suckers to fasten themselves to a carp or a stickleback, and a tube for sucking

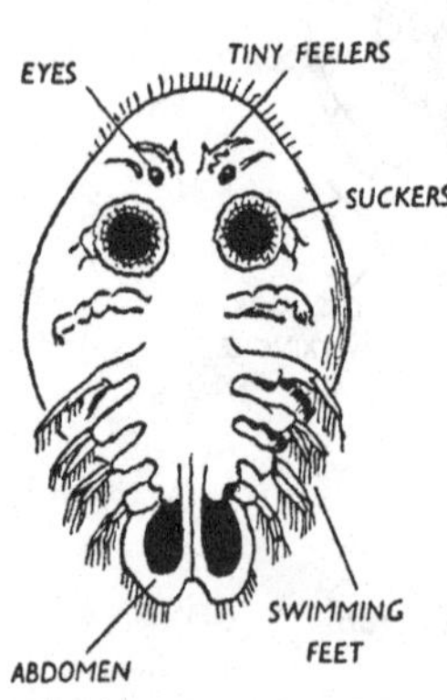

Fig. 38. *Carp louse. An external parasite on carp and sticklebacks*

the blood out of the fish. When they are not eating they can swim freely in the water by means of their four pairs of legs and their tail (Fig. 38). They lay their eggs on stones in the water.

Arachnids

Look at a spider and try to remember what you have learned about it in Book I. You will then be able to say what the characteristics of this family are. The body is divided into two parts: (1) the HEAD and THORAX which are joined together, (2) the ABDOMEN. They have eight EYES and eight LEGS. They breathe chiefly through BOOK LUNGS, but they have in addition a single breathing hole or SPIRACLE on the abdomen which leads into short air tubes or TRACHEAE. The smaller animals breathe through their skins.

The HARVESTMAN is found in gardens and has a small round body—which is not divided into two parts although it belongs to this family—and eight very long legs. It has no spinnerets, and it breathes through two spiracles which are near the bases of the fourth pair of legs. These lead into tracheae. It is a useful animal, because it eats APHIDES and small insect larvae. The eggs are laid in the soil and have not a case to protect them (Fig. 39*a*).

WATER SPIDERS are very similar to the garden spider, and even go to the surface of the water to breathe through their book lungs. Try to keep some in a large aquarium. Father water spider is bigger than the mother. These spiders spin silken threads in the water, up and down which they walk. Sometimes they make a bell-shaped dome of silk in the water. They weave a mass of threads together and fasten them to surrounding leaves. They then go to the surface of the water and take a bubble of air between their back legs. This they carry down into the water and release under the mass of threads, which gradually arches up to form a bell. In the breeding season the father water spider makes a bell adjoining that of the mother. He then breaks down the threads between the two bells to make one large one. You may be fortunate enough to see this happen in your aquarium. Mother water

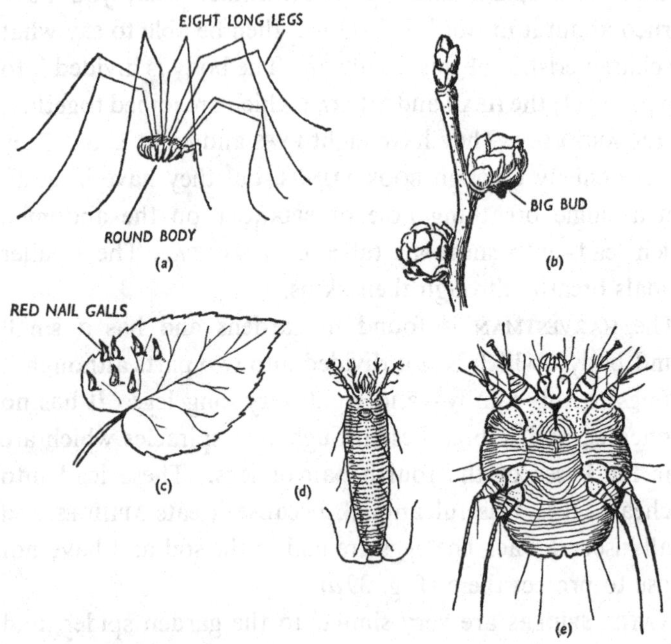

Fig. 39. *Arachnids. (a) Harvestman; (b) big bud of hazel; (c) nail galls on lime leaf; (d) nail gall mite (enlarged); (e) itch mite (enlarged)*

spider lays her EGGS at the top of the bell, where they are surrounded by air.

MITES and TICKS, which live as parasites on animals and on plants, also belong to this family. Small RED MITES live in ponds and swim about in the water, at first sucking the blood of water insects but later eating tiny crustaceans; other red mites are very harmful to fruit trees on which they live. The BIG BUD disease of currants and hazel trees is caused by a MITE that lives in the bud. The disease known as SCABIES is caused by a tiny mite called the ITCH MITE, which tunnels into a person's skin. The tunnels run parallel to the surface of the skin. As the mite tunnels she lays two to three eggs every day. It is this tunnelling that makes the skin irritate. Mites can walk out of a tunnel when they hatch out of the eggs. They can walk on to the bedclothes and on to another person. The tunnels are covered at first with a yellow blister which later forms a scab. Ointment must be rubbed into the tunnels to kill the eggs and the young ones. To do this the scabs must first be rubbed or scrubbed off. BLACKHEADS are tiny mites living in the skin. Mites also cause MANGE in dogs and ISLE OF WIGHT disease in bees. TICKS live on many animals and spread diseases. SHEEP TICKS are fairly big.

Myriapods

As these animals have been studied in some detail in Book I, we know that the chief characteristics of this group are: (1) the body is not divided into thorax and abdomen; (2) there are legs on every segment; (3) they breathe through air holes or SPIRACLES and AIR TUBES. MILLIPEDES and CENTIPEDES, which belong to this group, are described in Book I.

Insects

We have studied several insects, so we should know how to recognize them. The BODY is divided into three parts: (1) HEAD, (2) THORAX, (3) ABDOMEN. The thorax is divided into three segments and the abdomen is divided into ten segments. Nearly all insects have six LEGS which grow out of the thorax, and adult insects have four WINGS which also grow out of the second and third segments of the thorax. They have two COMPOUND EYES which are made up of hundreds of tiny little eyes. They breathe through SPIRACLES which lead into air tubes or TRACHEAE which penetrate all parts of the body.

Insects can be divided into two main groups: I, insects without wings, II, insects with wings.

I. INSECTS WITHOUT WINGS

To this group belong all the insects which never have any wings, and which change very little in structure when once

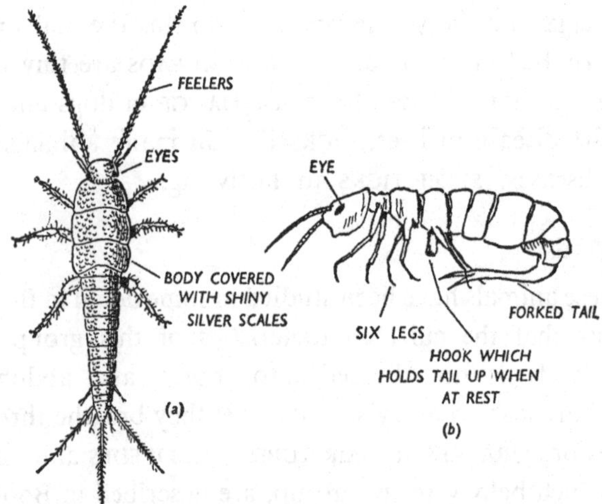

Fig. 40. *Insects without wings.* (*a*) *Silverfish;* (*b*) *springtail*

they have hatched. The SILVERFISH, which is commonly found in houses and which feeds on starchy foods, belongs to this group. Try to catch some of these insects at night-time when they are most active. Look at one with a lens and compare it with Fig. 40(*a*). If you look closely at stagnant water you may see some tiny animals called SPRINGTAILS jumping about on the top of the water. Each animal has a forked tail which is bent under its body. This is pressed on the water to make the springtail jump into the air (Fig. 40*b*). Young springtails are like their parents but they are smaller in size.

II. INSECTS WITH WINGS

This group can be sub-divided into two:

A. *Insects whose wings develop outside the body.* The young ones, which are called NYMPHS, are similar in form to the adults, but they have not any wings. As they grow older the wings gradually develop on the outside of the thorax.

B. *Insects whose wings develop inside the body.* The young ones are called LARVAE and they are not like their parents. There is always a PUPAL form when the larva changes into the adult.

A. *Insects whose wings develop outside the body.*

Several families belong to this group.

(1) *Cockroaches, stick and leaf insects, grasshoppers and crickets.* They have two pairs of wings, the first pair of which is harder and stronger than the second pair. These harder wings are laid flat over the body and protect the second pair, which are very delicate and are folded. COCKROACHES live in houses and bakeries, and eat any food that they can find. The female cockroach (Fig. 41*b*) carries her eggs in a case, which is about half an inch long and quarter of an inch wide (Fig. 41*c*). This case, which is hard, is fastened to the end of her body. She drags it along until she finds a suitable place to hide it. She cements it to the floor to protect it. If you keep cockroaches in school, give them any scraps of food to eat and somewhere to hide during the day-time, as they only come out to feed at night. CRICKETS and GRASSHOPPERS (Fig. 41*d*) are very similar, and have long back legs with

which they jump. Grasshoppers are flattened sideways and crickets are flattened from the dorsal to the ventral side. Both of them make peculiar noises by rubbing their legs on the

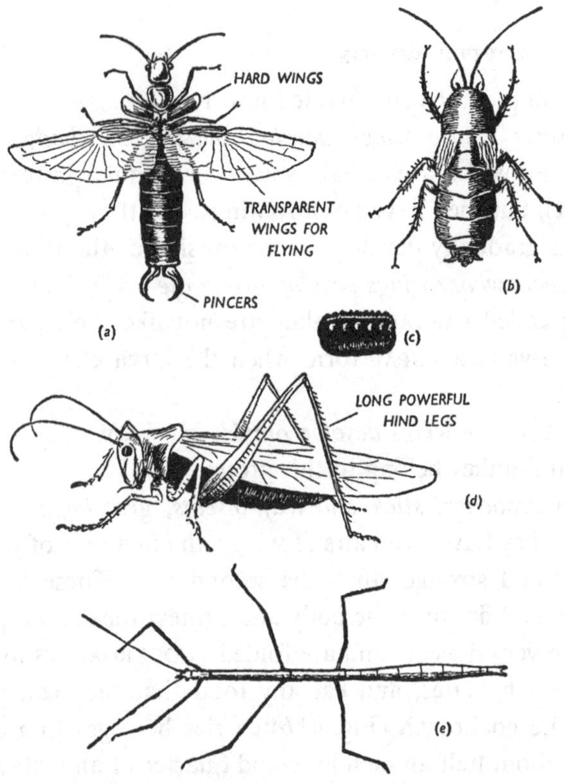

Fig. 41. *(a) Earwig with wings extended; (b) female cockroach; (c) cockroach egg case; (d) grasshopper; (e) stick insect*

rough sides of their wing covers. If you sit quietly on the grass where there are grasshoppers, when it is sunny you may see a grasshopper making this noise. They are vegetable feeders. The females lay eggs in little holes which they make in the soil. Crickets are found in houses and bakeries. STICK INSECTS also belong to this group. They come from warmer Mediterranean countries. They are very easily kept in school, as they eat privet leaves. They are very interesting insects to watch, as all of them lay eggs which hatch into tiny insects like their parents. The adults have not any wings and look like pieces of twig.

(2) *Earwigs.* Look at an earwig closely and compare it with Fig. 41 (*a*). It has a pair of PINCERS at the end of its body. The fore-wings are small and fairly hard. Beneath these the hind-wings, which are very delicate and transparent, are folded like fans. These wings could be used for flight, but earwigs seldom fly. The mother lays her eggs in a small hole in the soil. She watches over them until they hatch and then takes care of the young ones. Earwigs feed at night on animal and vegetable matter and hide in any dry place during the daytime.

(3) *Bugs of all kinds.* The wings of this group of insects develop gradually. Look again at the WATER BOATMAN, the WATER SCORPION, the FROGHOPPER and the APHID, and you will see that all these animals have a special mouth which looks like a long beak. It is with this beak that they are able to pierce the skin and suck the sap or the blood out of plants and animals. POND SKATERS and WATER GNATS, which walk on the surface of the water in a pond, also belong to this group.

(4) *Lice.* These are parasites. Some live on mammals, sucking their blood, others, which live chiefly on birds, are called BITING LICE. The body louse lives amongst our clothes and pricks the skin of the body to suck the blood. Head lice live amongst the hair and fasten their eggs, which are called NITS, to the hair. Although body lice are very seldom seen

today, head lice are still found on many people (Fig. 42*c* and *d*).

(5) *Dragonflies.* We have already studied these in Book I. Young dragonflies, which are called NYMPHS, live in water, and climb up the stalks of plants and out of the water when they are fully developed.

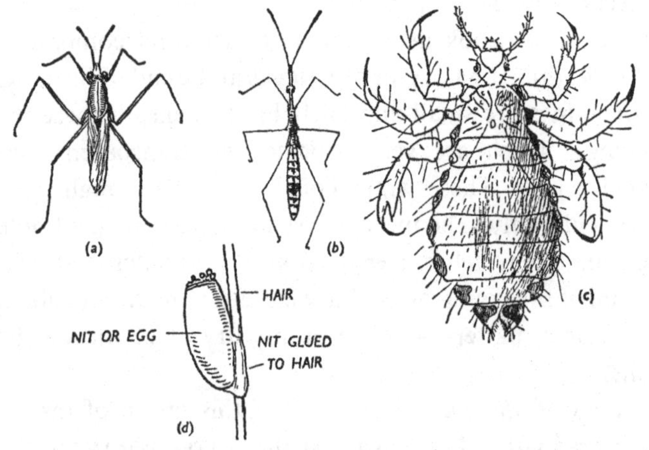

Fig. 42. *Bugs and lice.* (*a*) *Pond skater;* (*b*) *water gnat;* (*c*) *body louse;* (*d*) *egg or nit of head louse:* (*a*) *and* (*b*) *actual size,* (*c*) *and* (*d*) *much enlarged*

(6) *Mayflies.* These live in fresh water when they are young and fly in the air when fully grown. There are several kinds of mayfly (which vary in shape and in size), but all of them have three tails. Try to catch some when you go fishing. The young ones live for two to three years in the water eating decaying matter. Look at one closely and compare it with Fig. 43*b*. They swim very quickly and breathe through the gills which grow out of the side of the body. Adult mayflies do not feed and only live for a few hours. They may be seen flying above the surface of the water. During this short time the mother lays her eggs in the water.

B. *Insects whose wings develop inside their body*

There are four stages in the life history of all the insects that belong to this group: (*a*) EGGS; (*b*) LARVA, which is the eating and growing stage; (*c*) PUPA or resting stage, when the larva gradually changes into the last stage; (*d*) ADULT. The insects that belong to this group can be divided into several families.

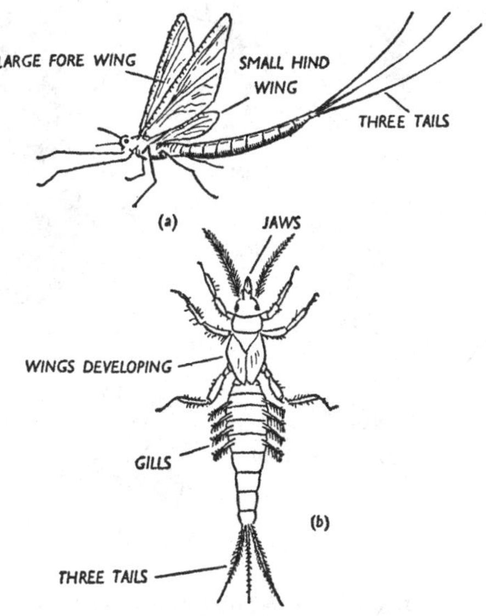

Fig. 43. *Mayfly.* (*a*) *Adult;* (*b*) *larva*

(1) *Beetles.* All adult beetles can be recognized by their wings. The first pair are very hard and horny and serve to protect the second pair, which are delicate and are used for flying. These wings are not so elaborately folded as are those of the earwig and grasshopper. There are many types of beetle which are found in water and on the land. Find as many as you can and try to identify them with the help of the books mentioned in Appendix C. Some beetles, such as

the CARNIVOROUS WATER BEETLE (see Book I), and the WHIRLIGIG BEETLE spend the whole of their lives in the water. Whirligig beetles are tiny black creatures which you may see whirling about very quickly on the top of the water. Some beetles are very useful in the garden as they eat harmful insects.

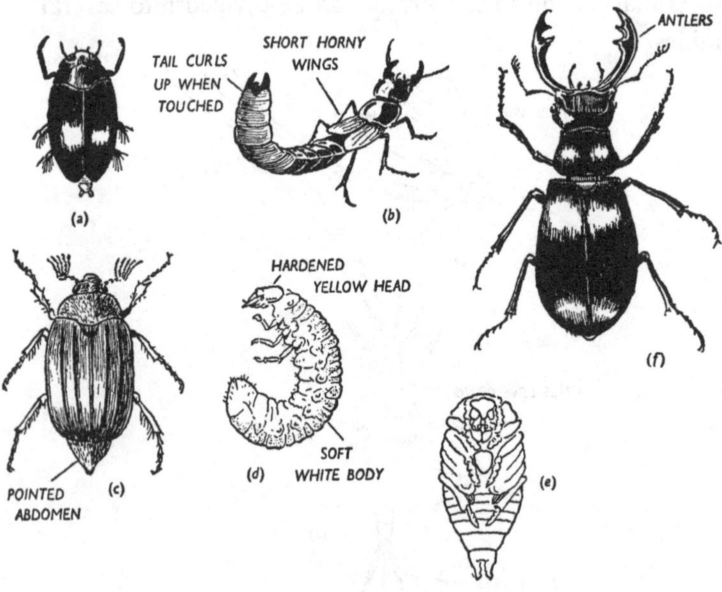

Fig. 44. *Beetles.* (*a*) *Whirligig;* (*b*) *cocktail;* (*c*) *cockchafer;* (*d*) *cockchafer larva;* (*e*) *cockchafer pupa;* (*f*) *stag*

LADYBIRDS and their larvae eat green-flies (see Book I). Many GROUND BEETLES and COCKTAIL BEETLES (Fig. 44*b*) are also useful. The green TIGER BEETLE, which is over half an inch in length, also hunts for harmful insects. Its larva lives in a small vertical burrow in the soil, with its head at the surface, ready to seize any prey that comes near. The larvae of GLOW-WORMS eat snails and slugs. The adult female is wingless, but she emits a light at night to enable the winged male to find her.

Many beetles are harmful because they destroy plants. CLICK BEETLE larvae eat roots (see Book I). COCKCHAFER BEETLES are very harmful, as they eat the leaves of forest trees during their short adult life of six weeks. The female lays its eggs in two or three clusters, with ten to fifteen eggs in each cluster. These hatch out into larvae which have a hard yellow head and a soft, white, fat body (Fig. 44*d*). These larvae may be so numerous in a field that they may completely destroy a field of hay by eating the roots of the plants. The larva changes into a pupa when it is three years old, and then it changes into a beetle. The larvae of the STAG BEETLE (Fig. 44*f*) are very similar to those of the cockchafer beetle. They live for about four years in the decaying wood of trees, and then pupate and change into beetles which live on the sweet juice that is given out by the oak tree. Stag beetles are very common in the southern part of England, especially in the Home Counties. The male stag beetle may be 2 inches long and can easily be recognized by its 'antlers', which are used to fight other male beetles. The females do not have 'antlers'. BARK BEETLES bite through the bark of a tree and then make little horizontal tunnels in the wood just under the bark (Fig. 45). WEEVILS are beetles that have long SNOUTS. There are about 25,000 kinds of weevil. The larvae have not any legs as they are surrounded by their food and do not have to hunt for it. They feed on peas, hazel nuts, blossoms of fruit trees, stored wheat, turnips, potatoes and many other plants, and cause much damage. Sometimes you may see many tiny holes in radish leaves. These are caused by very tiny FLEA BEETLES, which jump away like fleas when you go near to them.

(2) *Lace wing flies*. These are the most common insects that belong to this group. The adult has two pairs of wings

which are almost the same size and which are held over the back of the body, slanting like the roof of a house (Fig. 46*a*). The wings, which are greenish in colour, are longer than the

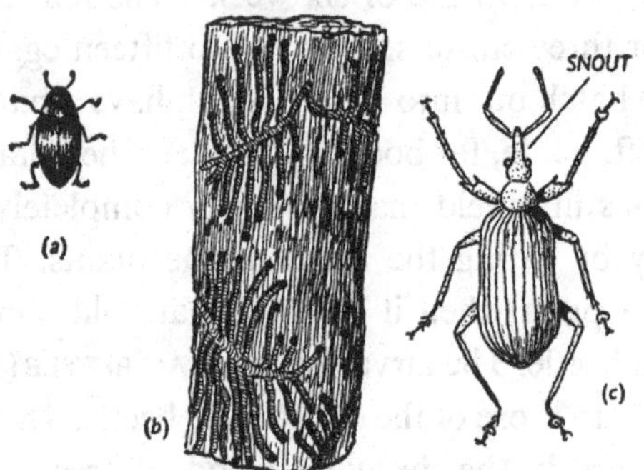

Fig. 45. *Beetle family.* (*a*) *Bark beetle;* (*b*) *tunnels made in wood by bark beetle;* (*c*) *weevil*

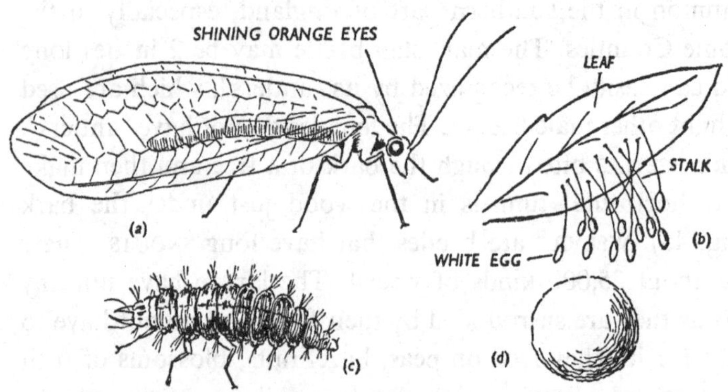

Fig. 46. *Lace wing fly.* (*a*) *Adult;* (*b*) *eggs;* (*c*) *larva;* (*d*) *cocoon containing pupa*

green body. The eyes are very big and are bright orange in colour. The feelers are very long. The EGGS are attached to leaves by long hair-like stalks. The dark brown LARVA, which has six legs, eats aphides (Fig. 46*c*). When fully grown the larva makes a round silken cocoon around itself and changes into the adult fly, which also eats aphides.

(3) *Caddis flies.* They have hairy wings and belong to another family of insects that have a complete life cycle.

(4) *The family of butterflies and moths.* They can be recognized by the tiny scales which cover their two pairs of wings and so give the wings their beautiful markings and colour.

As caddis flies, butterflies and moths are described in detail in Book I they are only mentioned here.

(5) *The fly family.* It is very easy to recognize, as all adult flies have only one pair of wings. The larva, which is called a MAGGOT, does not have any legs. We have studied the CRANE-FLY, the HOVER-FLY and the GNAT. The commonest fly is, of course, the HOUSE-FLY, which lays its eggs in rubbish or in dirty corners. The eggs hatch into maggots which live in the decaying matter. As they are surrounded by food they have not any legs or sense organs (Fig. 47). They change into chrysalides and finally into flies. House-flies are very harmful because they carry germs on the hairs on their bodies. These germs are put on to our food if they walk on it. When eating, the fly puts out a tube or PROBOSCIS from its mouth and ejects a substance on to the food to dissolve it. This liquid might contain germs. It then sucks up the dissolved food. MOSQUITOES, which are very similar to gnats, have a special mouth tube with which they can pierce the skin and suck blood. The female, which makes a shrill buzzing sound, sucks blood; whereas the male, which has bushy feelers, is quite harmless and sucks the nectar out of flowers. If a female mosquito should suck the blood of a patient who has malaria fever, it will pass the germs on to the next person whose blood it sucks. As the larvae and the pupae live in water and come to the

surface to breathe, many of them can be got rid of if the surface of the water is covered with oil. The BLOW-FLY lays its eggs on meat or on dead animals. One kind of fly lays its eggs on sheep. When the eggs hatch the maggots wriggle into the sheep's body and may cause the sheep to die. To prevent this all sheep are dipped in special liquids. The WARBLE-FLY

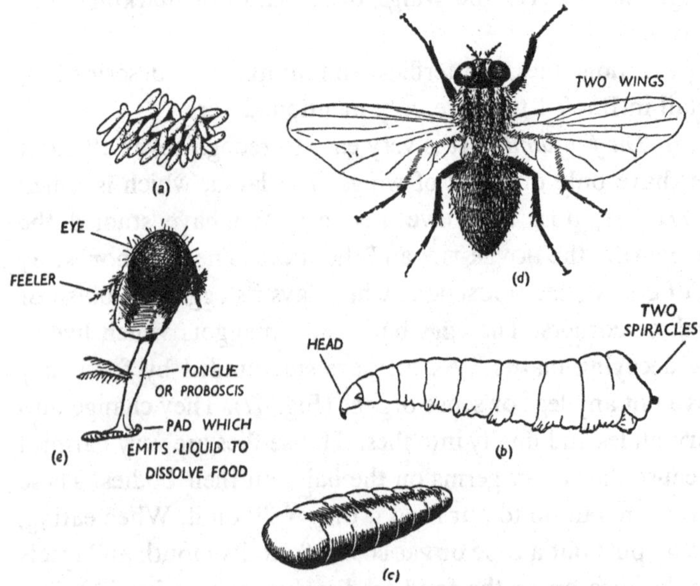

Fig. 47. *Stages in life history of house-fly:* (a) *cluster of eggs;* (b) *larva;* (c) *pupa;* (d) *adult fly;* (e) *side view of head* (enlarged)

lays its eggs on the legs of cattle. When the eggs hatch, the maggots wriggle through the animal's body, eating and growing as it moves about, and finally pupate beneath the skin of the host's back. These pupae are three-quarters of an inch long.

(6) *Fleas.* If you can find a hedgehog you will see many fleas running about amongst its spines. Catch one and look

at it. It has not any wings and its body is flattened sideways (Fig. 48). It has a piercing and sucking mouth tube to enable it to suck blood. The eggs drop to the ground and hatch into worm-like larvae which feed on any organic matter in the dust. They make cocoons which are covered with dust, inside which they change into fleas. They carry the germs of bubonic plague which caused the 'Plague of London'.

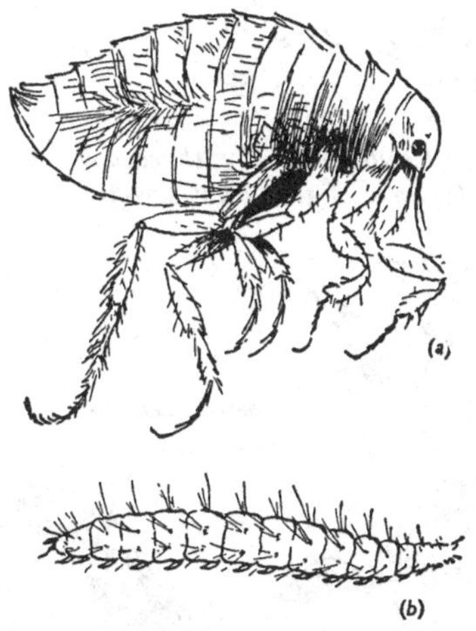

Fig. 48. *Flea. (a) Adult; (b) larva*

(7) *Ants, bees and wasps.* The insects which belong to this group have already been studied. Their chief characteristic is the wings. There are four transparent wings. The two wings on either side are held together by tiny hooks, so that they function as one wing. SAW-FLIES are interesting to watch. Their larvae look like caterpillars. You can always recognize them, however, because they have six to eight pairs of 'false legs' or PROLEGS on the abdomen, whereas caterpillars have either five or two pairs. The larvae of the hawthorn saw-fly

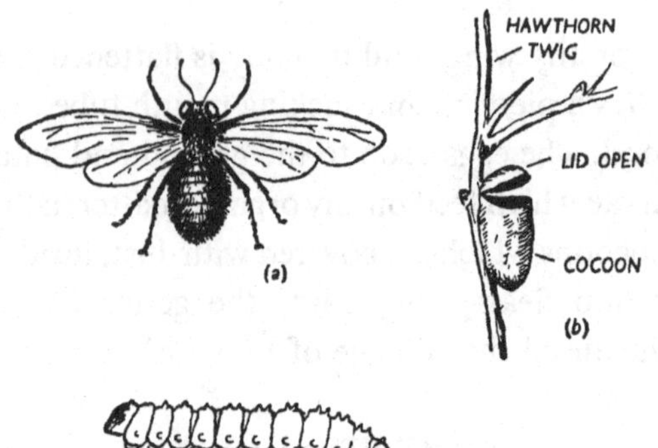

Fig. 49. *Saw-fly.* (*a*) *Hawthorn saw-fly;* (*b*) *cocoon of hawthorn saw-fly;* (*c*) *larva of gooseberry saw-fly*

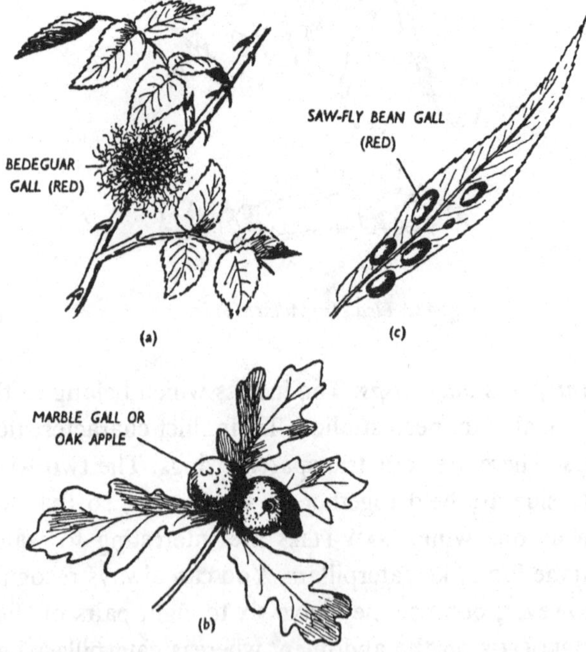

Fig. 50. *Galls.* (*a*) *Bedeguar gall on rose;* (*b*) *marble gall on oak;* (*c*) *saw-fly bean gall on willow*

make peculiar cocoons (Fig. 49*b*). When the pupa is ready to change into a saw-fly the lid comes off the cocoon to let it out. Some insects that belong to this family lay their eggs inside plants. A GALL forms round the young maggot. Try to find the red, hairy BEDEGUAR GALL on rose bushes, the MARBLE GALL on oak, and the BEAN GALL on willow (Fig. 50).

CHAPTER 5

ANIMALS: VERTEBRATES

The animals that have backbones can be divided into five main groups: (1) FISH, (2) AMPHIBIANS, (3) REPTILES, (4) BIRDS, (5) MAMMALS. We have already studied some animals that belong to each of these groups. Now we will try to find the characteristics of each group, and we will learn a little about the vertebrate animals that are not wild in this country.

Fish

Look at a fish and try to remember what you have already learned about it in Book I. Fishes can be easily recognized by the following characteristics: (1) they live in water and

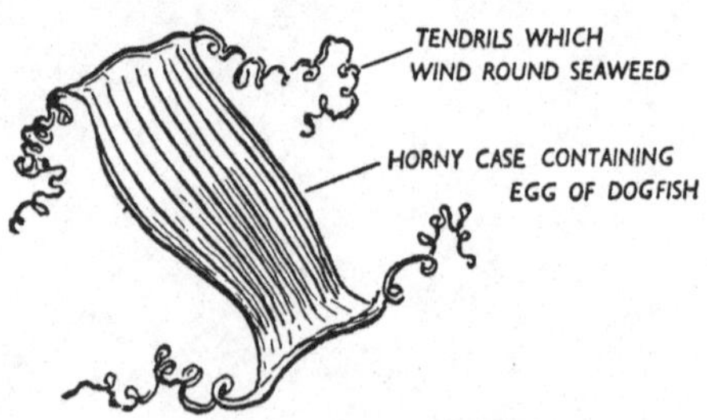

Fig. 51. *Mermaid's purse*

breathe through GILLS; (2) their bodies are covered with SCALES; (3) their eyes are not protected by eyelids; (4) they have FINS to help them to swim; (5) the sexes are separate and the eggs and the sperm are usually shed into the water where fertilization occurs. They often lay enormous numbers of eggs. The COD lays about five million eggs and the HERRING lays two million. Some fish, like the stickleback, lay only a few eggs, but these are guarded by the father. The DOGFISH and the RAY lay eggs singly. Each one is enclosed in a horny case and is fastened to seaweed. These cases are called 'mermaid's purses' (Fig. 51).

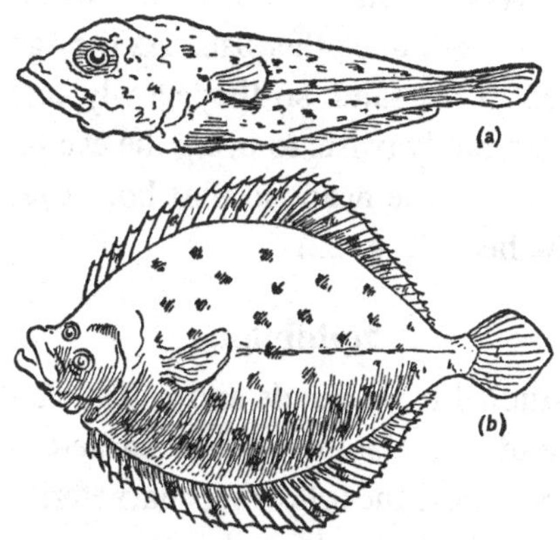

Fig. 52. *Flat fish. (a) Young fish which looks and swims like an ordinary fish; (b) adult fish (drawn to smaller scale). Turbot and brill lie on right side as above. Sole, dab, plaice and halibut lie on the left side*

The EEL that lives in our ponds and streams is a very interesting fish. When it is ready to SPAWN or lay its eggs, it swims down to the sea and thence across the Atlantic Ocean until it reaches the Sargasso Sea, which is near to the West Indies. Here it spawns, but what happens to it afterwards no one knows. It does not return. When the eggs hatch the young eels come back to the rivers and the streams, taking three years to do so. We know that this is so by the age of the fish that are found swimming on the surface at different distances from our shore.

The SALMON also has a peculiar life history. It spawns in the rivers. The young salmon remain in the rivers for about two years and then they swim to the sea. They remain for about two years in the sea, eating ravenously all the tiny animals that are floating about in the water. During this time they store fat in their bodies. Then they return to spawn. After spawning they may return to the sea again.

FLAT FISH are very peculiar animals. When the eggs hatch the tiny fish swim about like other fish. As they get older they like to lie buried in the sand, so they slowly turn on to one side. The skull gradually twists to bring the eye that would be underneath on top of the head, so that both eyes are on the same side of the head (Fig. 52).

Amphibians

In Book I we studied frogs, toads and newts. Look again at these animals and make a list of the characteristics of this group. You should note the following characteristics: (1) they have NAKED SKINS through which they can breathe; (2) they spend the first part of their lives living in the water like fish, and then their bodies change so that they can spend their adult life on land; (3) they have to return to the water to lay eggs.

Reptiles

The chief characteristics of this group are: (1) their bodies are covered with SCALES; (2) they breathe through LUNGS; (3) all of them, except snakes and slow-worms, have FOUR LIMBS; (4) most of them lay EGGS which hatch in the sun or in the heat of decaying matter where they are laid. We have already studied snakes, slow-worms and lizards which live in this country. Many reptiles live in hot countries and you can see them alive in zoological gardens. The living reptiles can be divided into four groups: (1) SNAKES; (2) LIZARDS and their relations; (3) TORTOISES and TURTLES, etc.; (4) CROCODILES and ALLIGATORS.

(1) *Snakes.* All snakes are similar in structure and vary chiefly in size and in colour. Most of them live in hot countries. You learned a little about the English snakes in Book I. Do you remember how to recognize them? They have long, thin bodies which are covered with scales. Unlike the other reptiles they do not have limbs. They have ribs throughout the whole length of the body which are free on the ventral side. The free end of each rib serves as a foot, each one ending in a scale which acts as the base of the foot. The eyelids are transparent and are joined. The lower jaws can stretch as the snake swallows prey that is fatter than its own body (see Book I).

(2) *Lizards.* They have long tails and four limbs. Unlike snakes, they have movable eyelids, fixed jaws, and ribs in the thorax only which are joined to the breast bone. A few lizards live in temperate countries, but most of them live in hot countries. GECKOS, MONITORS, IGUANAS, belong to the lizard family and are often seen in reptile houses in zoological gardens. The CHAMELEON, which lives in Africa, is about as big as an English lizard, but it has a very long tail which can be curled around the branch of a tree (Fig. 53). Its colour changes to match the colour of the leaf or twig on which it is standing, so that it is not easily seen, especially as it remains motionless for long periods. If an insect comes near to it the chameleon suddenly shoots out its long sticky tongue, which is club-shaped at the end. The insect sticks to the tongue and is passed into the mouth. The tongue can protrude 7 or 8 inches out of the mouth. The EYES of a chameleon are very peculiar. The eyelids (which are opaque)

are joined, but there is a small hole in the centre through which the chameleon can see. Each eye moves independently of the other one, so that it can look forwards with one eye and backwards with the other one at the same time. Its HANDS are also peculiar, as two fingers are opposed to the other three in the same manner that our thumb is opposed to our four fingers. This enables the animal to climb.

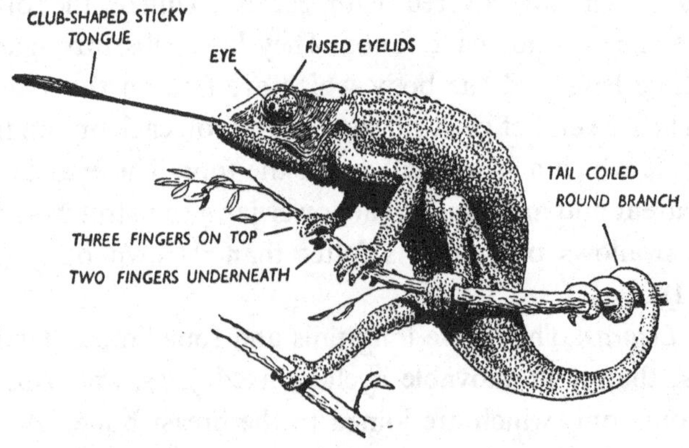

Fig. 53. *A chameleon*

(3) *The tortoise family*. These animals can quite easily be kept as pets, and TORTOISES, TURTLES and TERRAPINS can be seen in zoological gardens. These animals are recognized by the CASE OR CARAPACE which covers their body, and into which the head and legs can be withdrawn (Fig. 54*a*). This case is made of hard, bony scales or SCUTES, which are joined edge to edge. Around the edge of each scute you can see rings, which are GROWTH RINGS. Look at the case of a tortoise and you will see that the scutes are covered with horny scales that can be peeled off. It is from these horny scales that TORTOISESHELL is obtained. The backbone and some of the ribs are joined to the carapace. They have CLAWS on their

limbs to enable them to walk on land. Turtles, however, have limbs that are modified to form PADDLES for swimming (Fig. 54*b*). They have not any teeth, but they have hard jaws with which they eat green leaves. When the head is withdrawn into the case the neck bends into an S-shaped curve. All these animals lay eggs.

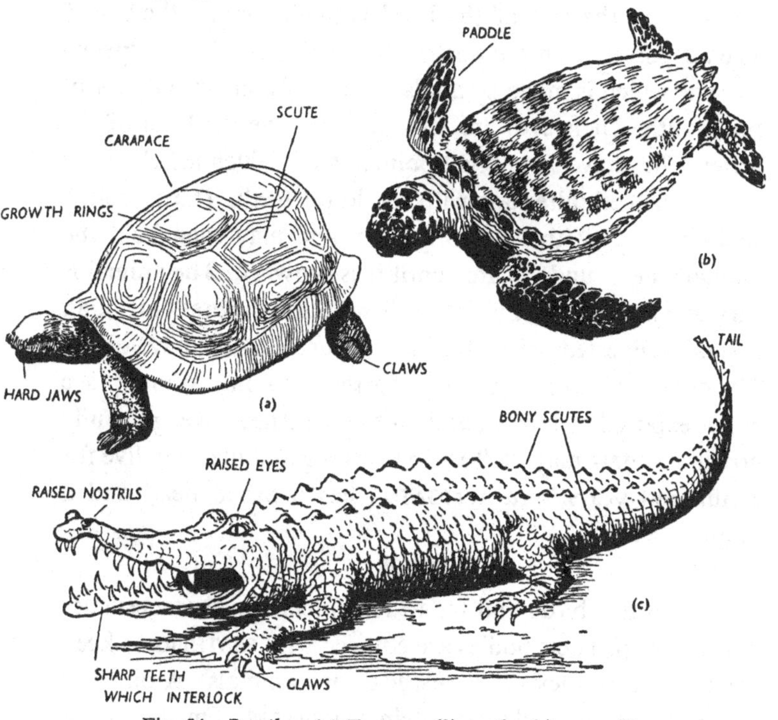

Fig. 54. *Reptiles.* (*a*) *Tortoise;* (*b*) *turtle;* (*c*) *crocodile*

(4) *The crocodile family.* The scales of the crocodiles and of the alligators, like those of the tortoise, are in the form of bony scutes which touch edge to edge. The scutes are covered with horny plates. These animals live chiefly in the water, and consequently their bodies are modified for this purpose. They

have four short legs for walking on land. They have a long, strong tail which is flattened sideways and which is used for swimming (Fig. 54*c*). They remain motionless in the water for long periods, almost totally submerged. Although they breathe with lungs, they can remain under water for some time, as the NOSTRILS are raised on the head and can be above water when the rest of the head is hidden under the water. The air passes from the nostrils to the throat, along a special passage which is made by a false palate of bone in the roof of the mouth. Their eyes are also raised above the level of the water. They have sixty-eight pointed teeth which interlock, as the upper teeth alternate with the lower teeth. The crocodile snaps its prey with its strong jaws, and may even hold the animal's head under water until it is drowned. The animal is very quickly digested; in fact, it is said that the first part of an animal is digested before the second part has been swallowed. The eggs are laid in a CLUTCH on the bank in a sunny place either exposed or buried, and mother watches the eggs. Fully grown animals may be 9 to 16 feet long and they can live for a hundred years. An ALLIGATOR has a broad head and a blunt nose.

Birds

Look at a few birds and try to remember the characteristics of this group. Their bodies are covered with FEATHERS to keep them warm, as they are WARM-BLOODED animals. A bird has very LIGHT BONES, which contain AIR SPACES. On the breast bone there is a KEEL to which the wing muscles are attached. The front limbs form WINGS. A bird is STREAMLINED in shape and it has not any external ears. It has three EYELIDS. In the eyes there are BONY PLATES which support the eyes against wind pressure. A bird does not have teeth; instead it has a HORNY BEAK for pecking food and a gizzard containing stones which grind it up. The two LEGS on which it walks are covered with SCALES. It has FOUR TOES (flightless birds have two or three toes), each of which ends in a CLAW. The LUNGS through which it breathes are connected with the air sacs in the bones. All birds lay EGGS, which are usually placed in a NEST. The eggs are hatched out by the warmth of the mother's body.

FLIGHTLESS BIRDS. There are a few birds which never fly. The PENGUIN (Fig. 55*a*), which comes from the Antarctic

Fig. 55. *Flightless birds.* (*a*) *Penguin;* (*b*) *emu*

regions, is a peculiar bird. It hobbles about on its two very short legs. Its wings are specially modified to form FLIPPERS for swimming, and it has webbed feet. It can glide in the snow on its chest. OSTRICHES, RHEAS, EMUS, CASSOWARIES and KIWIS do not fly. They are very heavy birds with long, strong legs and can run very quickly. They have not any wings or wing muscles, so they do not have a keel on the breast bone. Their feathers have not any barbules. They have very strong feet and have two (ostrich) or three (emu) toes (Fig. 55*b*).

Mammals

Look again at some of the mammals that you have studied and make a list of their chief characteristics. They all have HAIRS covering their bodies to keep them warm, as they are WARM-BLOODED. They breathe with LUNGS. They have FOUR LIMBS. There are TWO EYES, each of which is protected by TWO EYELIDS. There are TWO EXTERNAL EARS which protrude from the head. They have TWO SETS OF TEETH. The young are BORN ALIVE and they are fed on MILK which is secreted from special milk or MAMMARY glands, which may be in the thorax or in the abdomen.

Mammals are divided into several families, which can easily be recognized. You must remember what you have learned about the English mammals, because in this book we shall study the foreign ones that you can see in zoological gardens.

Peculiar mammals

In Australia there are several mammals that differ from those that are found in all other parts of the world.

The DUCK-BILLED PLATYPUS has a duck-like head, but it is covered with hair (Fig. 56*a*). The flat beak is covered with very sensitive skin. It has five toes on each foot, which are webbed. The web on the fore legs, however, is longer than the toes, but it is not fixed completely to them. When the animal is walking on land this skin can be folded back to expose the claws. It has a broad, flattened tail which helps it to swim. It catches small water animals, which are stored in pouches in its cheeks. The food is chewed later when the animal is drifting along in the water. Although these animals are mammals, they lay eggs. Just before the eggs are laid a special pouch develops on the abdomen of the mother.

Fig. 56. *Peculiar mammals.* (*a*) *Duck-billed platypus;* (*b*) *ant-eater;* (*c*) *rock wallaby*

The eggs are carefully put into the pouch, and when they hatch the young feed on milk from the mammary glands which are in the pouch.

The ANT-EATER also lays eggs. It has a very tiny tail and a very long snout (Fig. 56*b*). It has a long, thin tongue which is used for licking ants out of their nest, which it can tear open with its very long claws. Both of these egg-laying mammals belong to the same family, and are called MONOTREMES.

KANGAROOS, WALLABIES, WOMBATS and KOALA BEARS are also peculiar mammals, but they do not lay eggs. They all belong to one group and are called MARSUPIALS. The young are born alive, but they are imperfect and are extremely tiny. A baby kangaroo when it is born is only about an inch long. The mother has a permanent pouch into which the young ones are placed when they are born. The mammary glands are in the pouch, and the baby kangaroo holds on to the teat of the gland with its mouth. KANGAROOS and WALLABIES have short front limbs which have five fingers. The hind limbs are very long and are extremely powerful. There are five toes and the fourth one is the longest. A kangaroo can bound along the ground in leaps 7 or 8 feet long. The tail is only used for balancing when the kangaroo is resting. The KOALA BEAR eats the leaves of the eucalyptus tree. When a baby is big enough the mother carries it on her back. OPOSSUMS, which are also found in South America, have several babies, which cling on to their mother's back when they are big enough to leave the pouch.

Hoofed mammals or ungulates

If you think for a moment about horses, cows, sheep and pigs you will remember that you can recognize these animals by their HOOFS and by their TEETH. There is a space between the front and the back teeth. The grinding teeth are ridged to enable the animals to chew their food. The jaws move sideways as well as up and down when they are chewing. They are all vegetarians or HERBIVORES. Many of these animals go about in herds. The young ones are big and strong when they are born and can run about. They have TAILS and many of them have HORNS or ANTLERS (see Book I). The hoofed mammals are divided into two groups: (1) the ODD-TOED; (2) the EVEN-TOED.

Odd-toed. HORSES, DONKEYS, ZEBRAS and ASSES have one toe on each foot and they have not any horns. The RHINOCEROS, which lives in Africa or in Asia, has three toes on each foot. The body is covered with a tough skin which is thrown into folds, so that it looks like a suit of armour. The Indian rhinoceros has one horn (Fig. 57*a*) and the African rhinoceros has two horns which are one behind the other. These horns are made of a mass of hair-like structures which are stuck together and are fixed on a base of bone that is on the nose. TAPIRS, which live in South and Central America and Malaya, also have three toes. They live alone in marshy places and come out chiefly at night.

Even-toed. COWS and SHEEP have two toes, PIGS have four toes. The HIPPOPOTAMUS, which lives in Africa, has four toes. It has a thick, tough skin. It has not any horns. It spends most of its life in water, so its nostrils are on the surface of its head and can be closed when the animal is under water. It can remain under water for ten minutes, and it often walks on the bed of the river. PIGS and BOARS have very powerful snouts with which they can root up anything out of the ground. All the following even-toed mammals 'chew the cud'.

All animals similar to the DEER, ANTELOPE and SHEEP have two toes. They are very fast runners and they can climb precipitous heights. Antelope live chiefly in Europe and Africa

Fig. 57. *Hoofed mammals. (a) Indian rhinoceros; (b) Bactrian camel*

and deer are found throughout the world except Africa and Australia. CAMELS have very large feet with two toes, which prevent them from sinking into the sand (Fig. 57*b*). The camel stores fat in its hump. Look at the hump of a camel when you next visit a zoo. If it is full of fat it will be rigid and upright; but if there is not any fat in it, it will be flabby. The camel stores water in special pockets in its stomach. These pockets can be shut off from the stomach by special muscles. The BACTRIAN CAMEL has two humps and the DROMEDARY has only one hump. Camels are not found in America. The LLAMA, which does not have a hump and which has very long fur, really takes its place. The GIRAFFE lives in Africa. It has a very long neck which enables it to eat the leaves of trees. The giraffe has to stand with its legs apart if it wants to drink or eat anything on the ground. It can grow to a height of 18 feet. Although its neck is so long it consists of only seven bones, which is the same number as you have in your neck. It has small bony horns on its head. The colours of its body match the surroundings. The OKAPI, which lives in the Belgian Congo in Africa, always lives in pairs. It has a head like that of a giraffe, but its neck is shorter. Its body is dark brown, but there are a few light stripes which run horizontally round its chest and its legs.

OXEN are similar to cattle, only they are more heavily built animals. The BISON, which is found in America and the Ural Mountains in Asia, is a very big animal with heavy shoulders. It is a shaggy-looking animal at times when its fur peels off in patches. The YAK, which lives in the cold regions of Tibet, has long horns and very long fur which almost touches the ground.

ELEPHANTS are hoofed mammals, but they differ from all others in several ways. They have four nails on their front feet and three on their hind feet. Their skin has very few hairs and is very tough. The nostrils and the upper lip are drawn out to form a TRUNK, which can be used for breathing, for sucking up liquids and squirting them out again, for picking up large things such as tree trunks, or for picking up small things, which it does with the lips at the end of the trunk. Two front teeth project forward from the upper jaw and are called TUSKS. The elephant has two or three large grinding

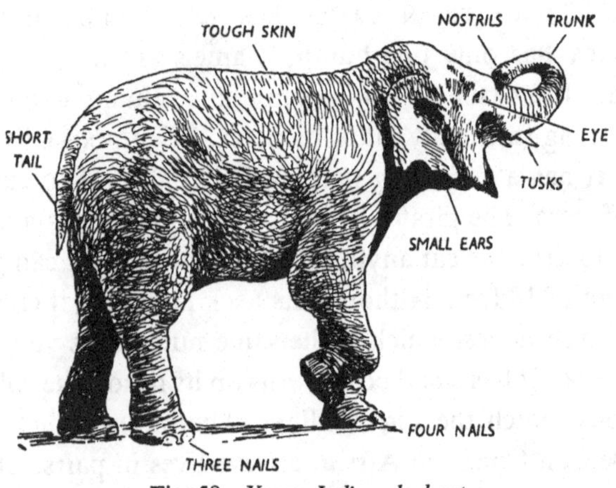

Fig. 58. *Young Indian elephant*

teeth on each side of the mouth in the top and bottom jaws. The eyes are fairly small (Fig. 58). The African elephant is bigger than the Indian one, and it has much larger ears. The ivory obtained from the African elephant is better than that from the Indian elephant. The latter are very easily trained to work or to act. Elephants go about in herds. The young, which are born singly, are well developed when they are born, so that they can join the herd for protection. An elephant is mature when it is forty years old and it can live for one hundred and fifty years. It can run at a speed of ten miles per hour.

Whales

It is not often that we see whales, but occasionally they are washed ashore. Built-up specimens of whales can be seen in the South Kensington Natural History Museum. Whales are the only mammals that live entirely in the sea. We know that whales are not fish because they breathe through lungs, their young are born alive, and they have not any scales. They have only a few hairs, which are round their noses. Their tails are flattened horizontally to allow them to come to the surface more easily to breathe. They have one or two nostrils or BLOW HOLES, which are on the top of the head. The lungs act as hydrostatic organs. This means that the whale can swim easily at different levels in the water because the amount of air in the lungs can be controlled. The deeper down in the water that it swims, the less air it has in its lungs. If it swims near to the surface there is more air in the lungs. Its front limbs form fin-like PADDLES, but it has not any hind limbs. The dorsal fin is fatty and not bony like that of a fish. As the whale's body is supported by the water, its bones are not very strong. If a whale is washed ashore the bones cannot support the weight of the body, and the chest is crushed by its own weight, so the whale cannot breathe. As it is a warm-blooded animal it has a thick layer of fat or BLUBBER under the skin.

There are two groups of whales:

(1) *Toothed whales.* They have large mouths and teeth, and feed on large animals. They have a single blow hole on the top of the head. As the whale nears the surface of the water to breathe it blows out air, which causes the water to shoot up like a fountain. The KILLER WHALE is 30 feet long. Some SPERM WHALES that live in the Tropics may be 82 feet long.

These whales are so called because they have oil in the cavities of the skull. This oil is called spermaceti. Whales are usually found in large numbers or SHOALS. PORPOISES, which are 6 to 8 feet long, are frequently seen around our coast, and sometimes swim up deep rivers in search of salmon.

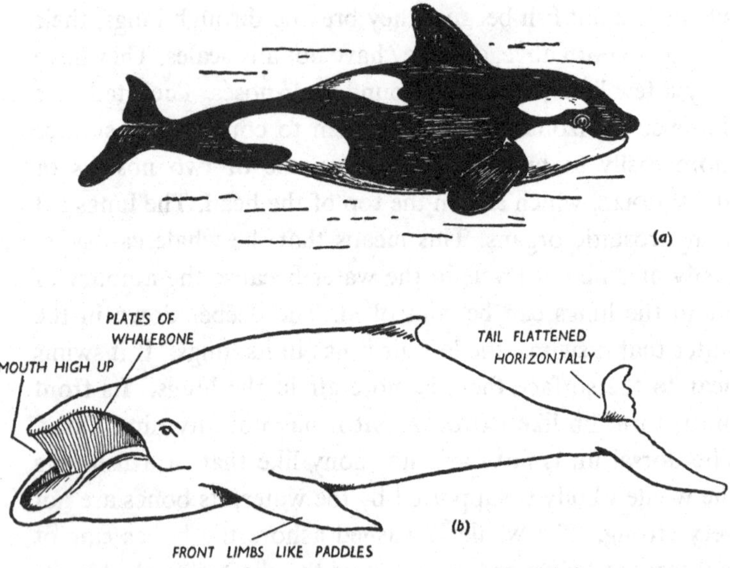

Fig. 59. *Whales.* *(a) Killer whale, a toothed whale; (b) lesser rorqual, a whalebone whale*

(2) *Whalebone whales.* They have not any teeth. Instead they have plates of WHALEBONE which hang down from the palate into the lower jaw. Each piece of whalebone is triangular in shape, the broader part being at the top. The lower edges are fringed. There may be as many as 370 plates, which may be 13 feet long at the base. With these plates the whales strain the water and eat all the tiny animals that are floating in it. The mouth is nearer to the top of the head than

it is in the toothed whales. RORQUALS and RIGHT WHALES, which may be 60 feet or more in length, belong to this group. Rorquals may stay under water for 12 hours, but the Greenland whale only stays under for just over an hour.

Fig. 60. *Beast of prey. Russian brown bear*

Beasts of prey

A few animals that belong to this group, such as the DOG, CAT, FOX, FERRET and WEASEL, live in this country. Beasts of prey live in all parts of the world and many of them can be seen in zoological gardens. They are easily recognized by their TEETH and by their CLAWS. The CANINE TEETH in these animals are long and pointed and are often called FANGS. All the teeth are sharply pointed and the points of the upper teeth fit into the depressions of the lower teeth. When the mouth is shut the upper jaw overlaps the lower one giving the animal a firmer grip on its prey. As all beasts of prey eat other animals, these sharp teeth and claws are necessary to enable them to catch their food and to tear it to pieces. These animals usually have large families, and the young are very helpless when they are born. This group can be divided into several families.

(1) *The cat family.* The animals that belong to this family have RETRACTILE CLAWS and ROUGH TONGUES. They usually go about alone or in families. We have already studied the cat in Book I. LIONS are found in Africa and Asia. They cannot climb. Their colour matches that of the sand on the ground. They roam about at night and roar when their stomach is full. Father lion has a very beautiful mane. TIGERS live in Asia and not in Africa. They are very fierce animals and can climb and swim. They have dark stripes on their bodies so that they cannot very easily be seen in the undergrowth. They are found in both hot and cold places. LEOPARDS, which live in Africa and in Asia have spotted skins which match the background when they are climbing trees. They are very fierce. The CHEETAH has very long legs and is the swiftest land animal in the world. The LYNX, JAGUAR, OCELOT and PUMA also belong to this family.

(2) *The dog family.* These animals do not have retractile claws. They hunt for their food in PACKS. They have five toes on their front and on their back feet, whereas cats have only four toes on their hind feet. Animals that belong to this group are similar to the dog and the fox which we have already studied. JACKALS, which live in Africa and the East, go about alone. The DINGO is the only beast of prey that lives in Australia.

(3) *The panda.* This belongs to another family. It is a great pet of all children because of its black and white colouring. Pandas live in the Himalayan Mountains, going as high as

12,000 feet. They live in trees and eat vegetable matter, eggs and insects.

(4) *Badgers, ferrets, stoats and weasels*, which we have already studied, belong to another family.

(5) *The bear family*. The palms and soles of the brown bears are naked, but the polar bears have hairy feet to enable them to walk on the ice. Polar bears live in the Arctic regions and eat seals. They spend the winter in holes which they dig in the snow but they do not go to sleep. Bears can often walk short distances on their hind legs. The SUN BEARS, which are fairly small, dark brown bears, have a white V on their chest. These bears are very good climbers. They are very fond of sweet things and will eat the honey out of bees' nests.

(6) *Seals and sea-lions*. These are aquatic animals. The hind limbs of the sea-lion are separate from the body, so

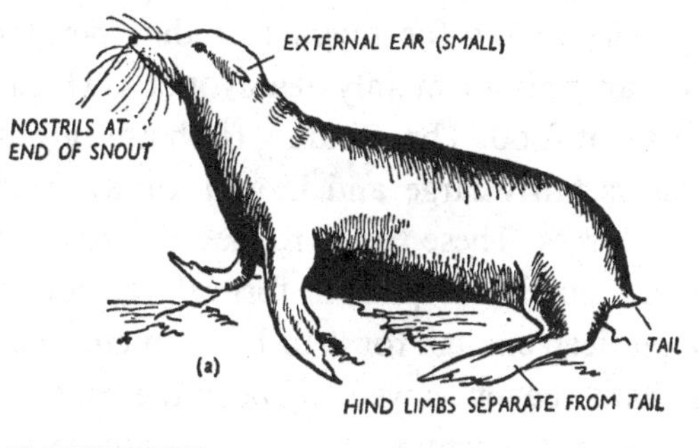

Fig. 61. *More beasts of prey.* (*a*) *Sea-lion;* (*b*) *seal*

that it is able to hop along on land. The hind limbs of the SEAL are joined to the tail, so that it cannot walk on land. The young ones remain for a time on land, hidden amongst the rocks. These animals have very small ears. The WALRUS, which is easily recognized by its enormous TUSKS, can walk on land.

Gnawing mammals or rodents

The animals belonging to this group have a space between their front and their back TEETH, but they are readily distinguished from hoofed mammals because they are all small animals and have CLAWS and not hoofs. RABBITS, HARES, RATS, MICE and SQUIRRELS, which we have studied, belong to this group. They have very large families, and the young are immature when they are born. The mother keeps the babies in a nest which she makes for them, and they develop very quickly. These animals are mainly HERBIVOROUS, which means that they eat plant food. The grinding teeth are ridged.

The BEAVER is fairly large and is well known because it gnaws and fells trees. These trees are placed across a stream to form a dam, so that a pool is formed. It does this to make the water deep enough for it to have an entrance to its home in the bank, below as well as above the water, so that it can escape from its enemies. PORCUPINES are peculiar because they are covered with long spines.

Insect-eaters or insectivores

HEDGEHOGS, MOLES and SHREWS, which we have studied, belong to this group. They are small animals with very tiny pointed TEETH all round the top and the bottom jaws. They have pointed SNOUTS. All have CLAWS.

Bats

As we have already looked at a bat we know that these animals are recognized by their WINGS. There are many kinds of bats. The VAMPIRE BAT, which sucks the blood of animals, is very small. The fruit-eating bats of the Old World are very big.

Primates

All the animals that belong to this group live in countries where the climate is hot. We can, however, always look at these animals in zoological gardens. The chief characteristics of this group are: (1) they have NAILS on their five FINGERS and five TOES that are similar to ours; (2) they have GRASPING HANDS and often GRASPING FEET, as their big toes are movable; (3) their EYES are at the front of their head; (4) they have well-developed BRAINS and can think and reason things out; (5) they are chiefly fruit eaters; (6) they usually have only one baby at a time.

American monkeys. These differ from Old-World monkeys because they have tails that can be used for climbing. Their nostrils are usually wide apart and point outwards. They have not any pouches in their cheeks for storing food. The MARMOSET has a very long tail which is not used for climbing. The SPIDER MONKEY has very long arms and a long, prehensile tail. The WOOLLY MONKEY has short, woolly, black hair. It has a prehensile tail which has a naked tip. The CAPUCHIN, HOWLING and SQUIRREL MONKEYS also belong to this group.

Old-World monkeys. This group can be divided into the MONKEYS and the MAN-LIKE APES. They have nostrils that are closer together than those of the American monkeys, and that point downwards. They often have bright pink patches under their tails which make them look very ugly. Many of them have cheek pouches for storing food.

Fig. 62. *Primates.* (*a*) *Rhesus monkey, the common, little monkey of the zoos;* (*b*) *chimpanzee;* (*c*) *orang-utan*

The MACAQUES have very nasty-looking bare patches. BABOONS are recognized by their pointed, dog-like noses. The MANDRILL has a very ugly face. Its nose is red and it has blue ridges on its cheeks.

The MAN-LIKE APES do not have tails. They have very long arms and walk in a semi-upright position, using the back of the hands when they are walking. They have loud voices. The GIBBON is the smallest ape and it has a very simple brain. It lives in the trees in Assam and Burma. GORILLAS live in Africa. A gorilla can grow to a height of 6 feet. It has massive shoulders and a broad chest. It has the strength of five or six men. Its arms are very long and strong, but its legs are fairly short. The hands and feet, too, are rather small. Its face is black and hairy and its nostrils are very wide. It has rather small ears. One male has several wives. At night the wives and children sleep on a platform which they build in the trees. The father stays on the ground on guard. During the day-time they wander about together. They are fruit-eaters. They look very sullen and very fierce and cannot be tamed. They can reason things out for themselves. CHIMPANZEES live in Africa. They are more intelligent than any other animal. They are very lively and playful, especially during the first seven years of their lives. They are easily tamed. The face is bare and the ears are large. The coat is black and glossy. The arms are very long. The ORANG-UTAN lives in Borneo and Sumatra. It is rather an ugly-looking animal, with reddish-brown skin and long, ragged, auburn hair. It has very long arms and a large, round abdomen. Its ears are very small. It lives chiefly in the trees and it is not as intelligent as the gorilla and the chimpanzee. Human beings are PRIMATES.

CHAPTER 6

A STUDY OF THE SOIL

We have studied the external appearance and the mode of life of many animals and plants, and we must now try to find out how they live.

As all living things depend on plants for their existence, we will begin by studying the life activities that are carried out by a plant. Before we can do this, however, we should know something about the soil in which the roots of a plant grow.

Look at a cutting through the earth, such as you will see in a pit or a quarry, or in a railway or road cutting. You will see that the top layer of the earth, which is called SOIL, is usually not very deep and is darker in colour than the soil that is below it, because it contains decaying animal and vegetable matter which is called HUMUS. The lighter-coloured soil below is called SUBSOIL and contains no humus. Soil and subsoil are made from rocks that have been broken up (Fig. 63).

Formation of soil

Rocks are broken up in many ways.

Water. As rain falls it dissolves some of the carbon dioxide that is in the air, and weak carbonic acid is formed which dissolves the chalk in the rocks. We know that water expands when it freezes. Where the rocks are exposed, the water in tiny cracks expands as it freezes, so making the narrow cracks much wider. Broken pieces of rock are washed away by the rain, and are broken into smaller pieces as they are hurled to the valley below. Rocks along river banks or by the sea are gradually worn away or ERODED by the water which dashes against them. The soil that is so made is carried away by the water, and is deposited elsewhere. Some places in England, such as Southport in Lancashire, which were once covered by the sea, are gradually being left farther

Fig. 63. *Cutting through soil*

behind by the sea as more and more soil is deposited on the shore.

Wind may blow away loose soil during very dry weather, and leave the rocks exposed.

Temperature. In some countries there is a great difference between the day and the night temperatures. The rocks expand in the heat during the day-time, and contract when it is much cooler at night. This alternate expansion and contraction of the exposed rocks causes them to crack.

Plants. Roots of plants may penetrate into the cracks of rocks which are widened as the roots grow. Have you seen pale green, or orange-coloured leaf-like growths on the surface of rocks? These little plants, called LICHENS (see page 36), are able to grow on the bare surfaces of rocks because their tiny little root-like growths give out a weak acid which dissolves the rock.

Types of soil

The chief types of soil are CLAY, SAND, LOAM, CHALK and PEAT. Soil maps of the country are now compiled so that

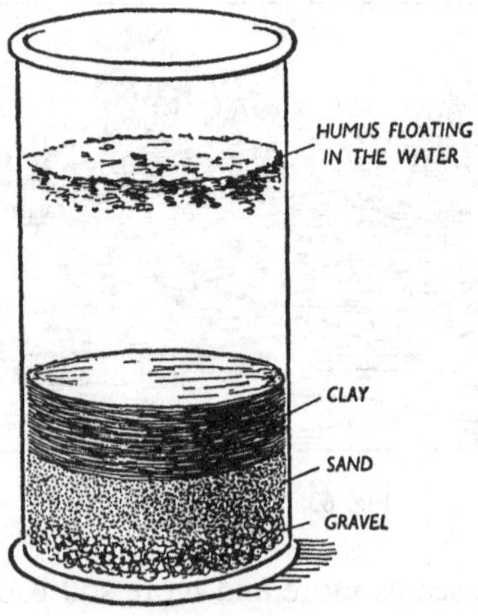

Fig. 64. *Experiment to show the constituents of soil*

a farmer can know the best crops to grow on his land, and also how to improve his soil. The following experiment will give you a rough idea of the type of soil that you have in your garden.

Experiment 1. Put a little damp soil into a 2 lb. jam jar and fill the jar with water. Mix it up thoroughly and then leave the soil to settle. Sand falls to the bottom first because the particles of sand are big and heavy. Clay particles which are fine settle on top of the sand, and humus floats on the top of the water (Fig. 64).

Before you can carry out some experiments on soil, you must get a quantity of each type of soil, and leave it to dry. Do not dry it in an oven. When the soil is dry, pound it up with a pestle and mortar and sift it.

Clay

Look at particles of sand and of clay under a microscope and you will see that sand particles are much larger than clay

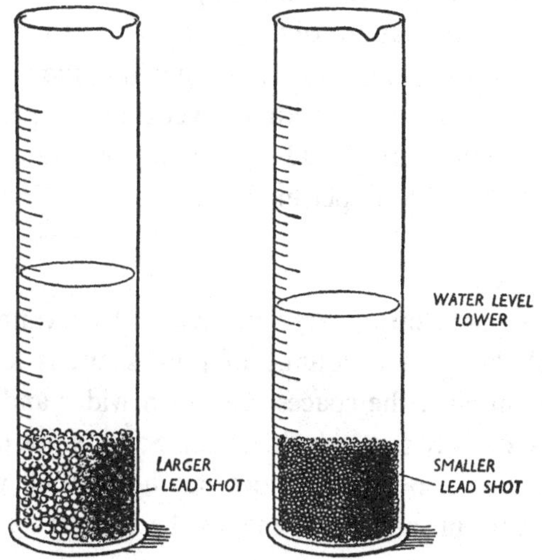

Fig. 65. *Experiment to demonstrate the meaning of 'pore space', and to show that the smaller the particles the greater the pore space*

particles. Between the particles of soil there is always a space called the PORE SPACE.

Experiment 2. To illustrate this put small lead shot into a measuring cylinder up to the 20 c.c. mark, and pour 30 c.c. of water on to it. The total is 50 c.c., but you will see that the reading is less than this. Some of the water has filled up the pore space between the lead shot (Fig. 65).

Experiment 3. Carry out Experiment 2 using different-sized lead shot in different measuring cylinders. You will find that the smaller the particles the greater the pore space.

Although the pore space is greater in clay than it is in sand, the spaces are very narrow and the water cannot run through the soil quickly. In addition, the particles of clay are able to adsorb and hold water. So the pore space which should be filled with air becomes filled with water, and the ground becomes WATERLOGGED. Waterlogged soil is bad for plants, because the roots cannot obtain oxygen and so they die. The water which is held in the pore space makes the clay soil very COLD and HEAVY. If you walk on wet clay and press it down it becomes PUDDLED. Puddled clay will not allow water to pass through it (see Experiment 4).

Sand

Sand is made up of large particles which cannot adsorb water. Although the volume of pore space is less in sand than it is in clay, the spaces are much wider and the water runs very quickly through the sand. Sand is not a good soil for some plants because it does not hold much water. The lack of water in sandy soils makes the soil WARM and LIGHT to work.

Loam

Loam is a mixture of sand, clay and humus in varying proportions. When animal and vegetable matter decays to form humus, mineral salts and nitrates are put back into the soil.

Peat

Peat is humus that has been formed under water. It is a very acid soil as acid is formed when the plants or animals decay.

Experiment 4. The following experiment demonstrates very clearly the capacity of different soils to hold water. Take four hollow glass tubes about 8 inches long and 1 inch diameter, with a lip at one end. On to this end of each tube tie a piece of muslin. Using dry, sifted soil, half fill each tube with different kinds of soil: (*a*) pure sand, (*b*) loam, (*c*) clay. Into a fourth tube put puddled clay. You can do this by pressing

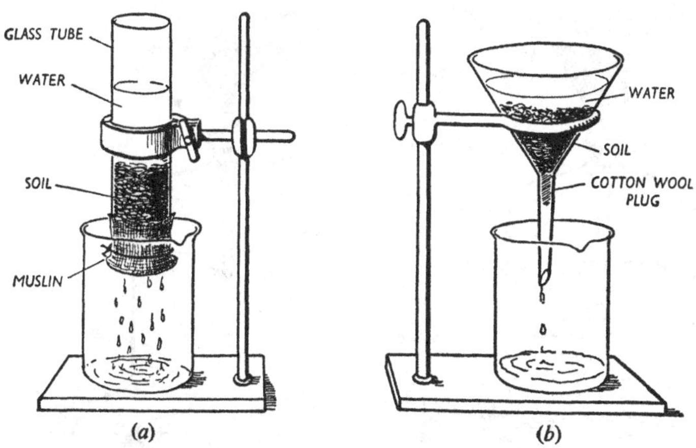

Fig. 66. *Two types of apparatus to demonstrate the water-holding capacity of soils*

wet clay into the tube from the bottom. Then tie muslin over the bottom of the tube. Place each tube into a clamp and place a jar under each one (Fig. 66*a*). If you cannot obtain these tubes, you may use funnels with the necks plugged with cotton wool (Fig. 66*b*). Let four children simultaneously pour the same amount of water into each tube and watch the tubes. You will notice that the water runs very quickly through the sand, fairly quickly through the loam if it is a sandy loam, not quite so quickly through a clay loam, and very slowly through the clay. The water remains on top of the puddled clay as it cannot sink into it.

Experiment 5. You can find out which type of soil holds the most water in the following way. Repeat Experiment 4 but place the end of each tube into a measuring cylinder which is just wide enough to hold the tube (see Fig. 70). This will reduce the evaporation of water. Pour the same known

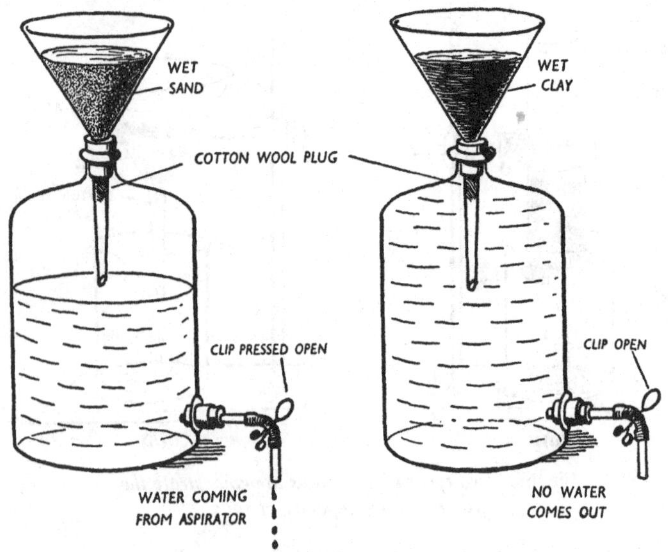

Fig. 67. *Apparatus to show the types of soil that allow air to pass through them*

quantity of water into each tube, and note the volume of water in each measuring cylinder when water has stopped dripping from the soil. You can then calculate the amount of water that has been retained by each type of soil.

Experiment 6. This experiment will show you that air can pass through wet sand but not through wet clay. Set up the apparatus as shown in Fig. 67. Place a plug of cotton-wool into each funnel and then put wet clay in one funnel and wet sand in the other one. Make quite sure that the apparatus is airtight round the bungs, and then open the clips. If air can pass through the soil, water will run out of the jar. You will notice that water runs only from the jar with the sand. This shows that the wet sand allows air to pass through but the wet clay does not.

Chalk

Chalky soils, which are very sticky in wet weather, should be mixed with loam or sand in order to yield good crops. Clay mixed with a fair amount of chalk is called MARL. If you live in a district where the soil is chalky you can carry out all the soil experiments mentioned, using your soil as well as sand and clay.

Important factors in the soil

Plants require water, air, food and the correct temperature, to grow well.

Water supply

The amount of water in the soil depends on several factors: (1) the CLIMATE determines the amount of rainfall; (2) the SLOPE OF THE LAND is important, as water runs away more quickly on sloping ground than it does on flat land; (3) the SUBSOIL too is important, because if the subsoil will not allow the water to drain away, then the soil will become waterlogged even if it is sand; (4) the TYPE OF SOIL is very important as clay soils hold more water than sandy soils.

When soil is apparently dry it still contains a certain amount of water which can be used by the plants. This water forms a thin film round each of the soil particles.

Experiment 7. This experiment will show you that apparently dry soil contains water. Put a little dry soil into a test-tube and gently heat it. You will notice that water vapour is given off and drops of water collect on the inside of the test-tube.

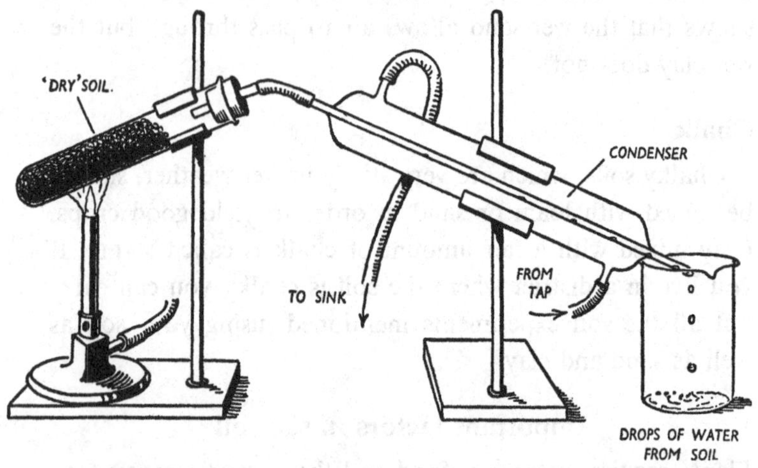

Fig. 68. *Experiment to show that 'dry' soil contains water*

Experiment 8. This experiment also shows that dry soil contains some water and here you can collect the water that has been driven off. Set up the apparatus as shown in Fig. 68. Gently warm the soil in the test-tube. Water vapour passes down the centre tube of the condenser. As this tube is surrounded by the cold-water jacket, the water vapour is changed into water which is collected in the beaker.

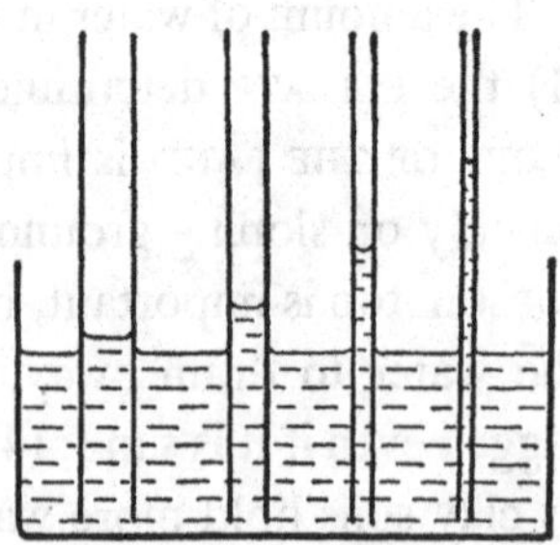

Fig. 69. *Experiment to show that water rises highest in the narrowest tube*

If the surface layer of soil becomes very dry, the water which is deep down in the soil may creep up through the pore spaces. Water creeps very easily in clay soil where the pore spaces are very narrow.

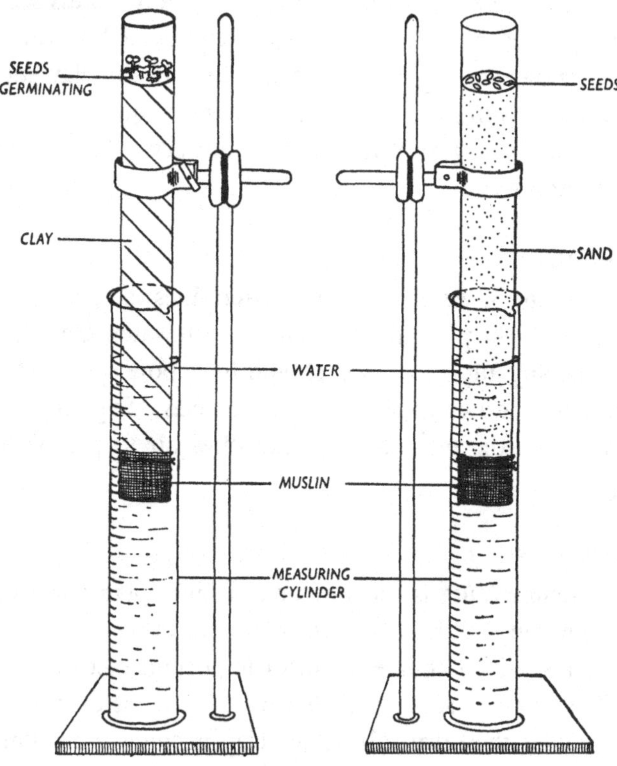

Fig. 70. ***Experiment to show that water creeps higher in clay than it does in sand***

Experiment 9. Place several narrow tubes of different bore into coloured water (Fig. 69), and you will notice that the water rises highest in the narrowest tube.

Experiment 10. To show that water creeps more quickly through the narrow pore spaces in clay than it does through the wider pore spaces in sand, take two tubes about 18 inches long, similar to those used in Experiment 4. Cover one end of each tube with muslin and then almost fill one with dry sand and the other one with dry clay. Place a few cress seeds on the top of each tube. Put the lower end of each tube into water as shown in Fig. 70. The cress seeds in the clay soon germinate, showing that the water has risen quickly in the clay. The seeds on top of the sand do not germinate as they have not any water.

Air supply

The amount of water in the soil determines the amount of air present. If the pore spaces are filled with water, then there will not be any air. Good soil should have pore spaces filled with air, with water round the soil particles. Experiment 6 will show you the types of soil that allow air to pass freely through.

Temperature

If the temperature is too low or too high, plants will not grow. The temperature depends on several things:

(1) CLIMATE is of course the most important factor.

(2) The TYPE OF SOIL. Soil that contains a lot of water is always colder than dry soil, hence clay is colder than sand. Plants growing in sandy soil are usually several weeks ahead of plants growing in clay soil if all other conditions are the same.

(3) The COLOUR OF THE SOIL. Humus makes soil dark. Dark soils absorb more heat than light soils, and so they are warmer.

(4) The SLOPE OF THE LAND. A southern slope gets direct rays from the sun, and so is warmer than land that slopes in any other direction or land that is flat.

Experiment 11. If you have a school garden, make plots of land that contain different kinds of soil. Set the same kinds of plant in each plot and notice the results. Also, if you have a greenhouse, set the same kinds of plant in the greenhouse and in the garden and notice the results.

Food supply

Plants obtain water and mineral salts from the soil. For a long time we have known that there are ten chief substances which a plant requires and without which a plant cannot grow and be healthy. These substances are carbon, hydrogen, oxygen, nitrogen, sulphur, phosphorus, potassium, calcium, magnesium and iron. A plant obtains carbon and oxygen from the air, and hydrogen and more oxygen from water. The remaining substances are dissolved in the soil water and enter the plant through its roots (see page 127). Recently it has been found that many more substances which appear in very minute quantities or traces in the soil are very important to some, if not to all plants. These substances are called TRACE ELEMENTS, the most important of which are boron, manganese, cobalt, silicon, zinc and copper. The absence of these substances causes certain diseases, for instance, lack of boron causes the centre of beet to rot. Experiments have been carried out to discover the use of each substance. These experiments are called CULTURE EXPERIMENTS. The following experiment is a simple culture experiment.

Experiment 12. Take six wide jars fitted with cork shives, sterilize them and rinse in distilled water. Pierce a hole through the middle of each shive and cut it as shown in the diagram. Make up a solution using distilled water which contains all the minerals necessary for the healthy growth of a plant, and then make other solutions with one mineral missing:

(1) COMPLETE solution. One quart distilled water. Enough saltpetre (potassium nitrate) to lie heaped up on a sixpence; and as much of each of the following as will lie heaped on a silver threepenny piece: Epsom salts (magnesium sulphate), powdered plaster of paris (calcium sulphate) and acid sodium phosphate. Put a rusty nail into the jar to supply the necessary iron.

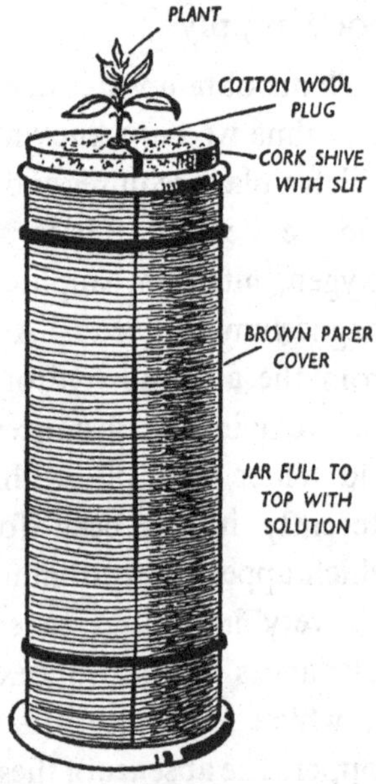

(2) For solution without NITROGEN omit the saltpetre and use potassium chloride.

(3) For solution without POTASH use sodium nitrate instead of saltpetre.

(4) For solution without PHOSPHATES omit the acid sodium phosphate.

(5) For solution without CALCIUM omit the calcium sulphate.

(6) For solution without IRON omit the rusty nail.

Fig. 71. *Water culture experiment. Diagram showing how each jar should be set up*

In Appendix B an alternative and more accurate method of making culture solutions is given.

Pour the solutions into the six bottles, and place similar seedlings through the holes in the shives, and hold them in position with cotton wool. (If you use seedlings that still have a fair amount of food in their seeds, you must remove the seeds.) Cover the jars with brown paper to exclude the light, and leave the jars in a strong light for several weeks

(Fig. 71) and notice the results. Make more of each solution than you want as you may have to fill up your jars. The jars must be aerated and the solution should be changed every month.

Improvement of soil

If soil is not in a good condition for growing plants it can be improved by cultural operations or by adding manures or fertilizers.

Cultural operations

These operations are necessary to prevent the soil from being too compact or too wet, and generally to improve the condition of the soil. DIGGING or PLOUGHING is a cultural operation in which the soil is turned over. FORKING loosens the soil. HOEING is done to get rid of small weeds and to widen the pore spaces to prevent the soil from losing water. ROLLING narrows the pore spaces and brings up the water for the young seedlings. Farmers often roll their ground during the spring if the weather is dry. If clay soil is left in RIDGES during the winter, the frost will break it up. RAKING or HARROWING gives the surface a fine tilth for young seeds. DRAINAGE is necessary in wet soil. Farmers usually dig ditches along the hedgerow to help in the drainage of the land. If the soil is very wet, drainage pipes may be laid throughout the field (Fig. 72).

Manuring

There are two kinds of manure which may be added to the soil:

(1) *Organic manures*. These are horse manure, cow manure, poultry manure, bone meal, dried blood, etc. All these

manures are expensive to buy, so many gardeners make their own manure in a COMPOST HEAP. Any garden refuse can be used except diseased plants, hard and woody plants, perennial weeds such as couch grass (see Book I) or weeds that have seeds. Place a layer of rubble at the bottom for drainage and then make the heap of alternate layers of 12 to 18 inches of

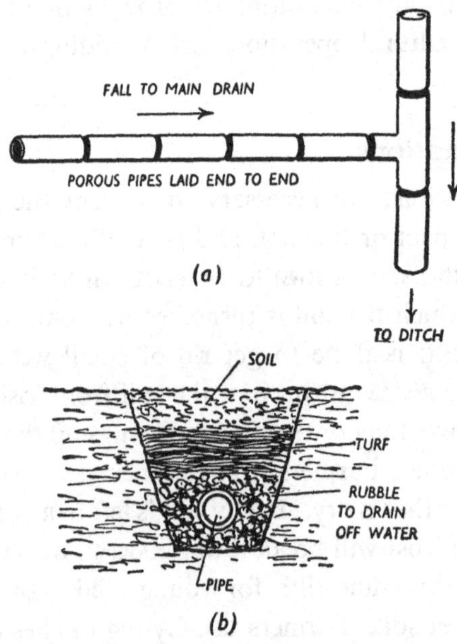

Fig. 72. *Field drainage: (a) pipes seen from above; (b) section showing method of laying pipes*

organic matter and 1 inch of soil, and dust each layer with lime. Tread each layer down firmly, and cover the heap to keep out the rain. The heap will decay in 6 to 16 weeks and can then be used as manure. Farmers sow clover which they plough into the soil as green manure. This increases the amount of nitrates in the soil (see page 32).

(2) *Inorganic manures.* They are usually called FERTILIZERS and are sold in all seed shops. You must be careful to use the correct amount of fertilizer and to use each fertilizer on the right crops and at the stipulated time of the year. There are three main groups of fertilizers: (*a*) NITRATES, such as sulphate of ammonia, nitrate of soda, nitrate of potash and soot. Nitrates are used to improve plant growth and leaf formation. (*b*) POTASH, such as sulphate of potash, which is used chiefly to develop root crops. (*c*) PHOSPHATES, such as superphosphate and basic slag. Phosphates improve the fruit-bearing of a plant as well as the root and leaf growth.

Experiment 13. This experiment will demonstrate the use of various fertilizers. Use five flower pots of not less than 10 inches diameter. Use 10 g. superphosphate, 2 g. sulphate of potash and 4 g. of nitrate of soda to each square foot of surface area. Use only one-third of the nitrate of soda at first and apply the remaining two-thirds at intervals of one month. In pot 1 put no fertilizer, in pot 2 put all fertilizers, in pot 3 omit superphosphate, in pot 4 omit sulphate of potash and in pot 5 omit nitrate of soda. Use oats and mustard seeds, or any other kind, and note the results.

LIME may also be added to improve the fertility of the soil.

(1) Lime causes the small particles of clay to stick together in clusters, or FLOCCULATE, and then water can run more easily through it.

Experiment 14. Set up two tubes as in Experiment 4, putting clay in one, and clay and lime that have been thoroughly mixed in the other. Add water to both and watch them. Water runs more quickly through the lime and clay than it does through the clay.

Experiment 15. This experiment will show that lime makes clay flocculate. Take two jam jars and put powdered clay into one and powdered clay and lime into the other. Pour water on to each, stir vigorously and then allow to settle. Watch the jars closely and you will notice that the particles in the jar containing lime settle first, which shows that they are heavier as they are stuck together (Fig. 73).

(2) When organic matter rots, acid is formed which makes the soil sour. Plants will not grow in sour soil. Lime kills or neutralizes this acid. The following experiments will show you whether your soil has too much acid or too much lime in it.

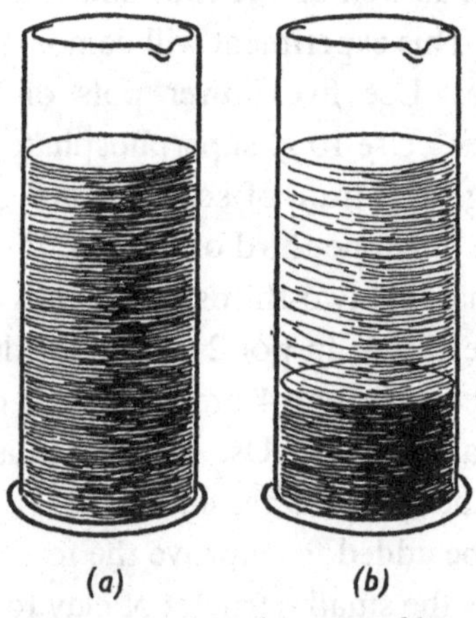

Fig. 73. *Diagram to show the action of lime on clay: (a) no lime added; (b) lime added*

Experiment 16. Place a piece of blue litmus paper in vinegar or hydrochloric acid and it will turn red. Place red litmus paper on lime and it will turn blue. Now mix some soil with a little distilled water. Place blue litmus paper in the liquid and if it turns red your soil is sour.

Experiment 17. For this experiment you need a B.D.H. soil indicator, special porcelain boats and capillary tube pipettes which can be bought from a firm selling scientific equipment. Place a little soil in the larger end of the boat and add a little B.D.H. liquid. Stir it up and then pour off the liquid into the small end of the boat. Pipette some of the liquid up into the capillary tube and compare the colour with that in the capillary tubes of the B.D.H. soil indicator. If the colour is a number

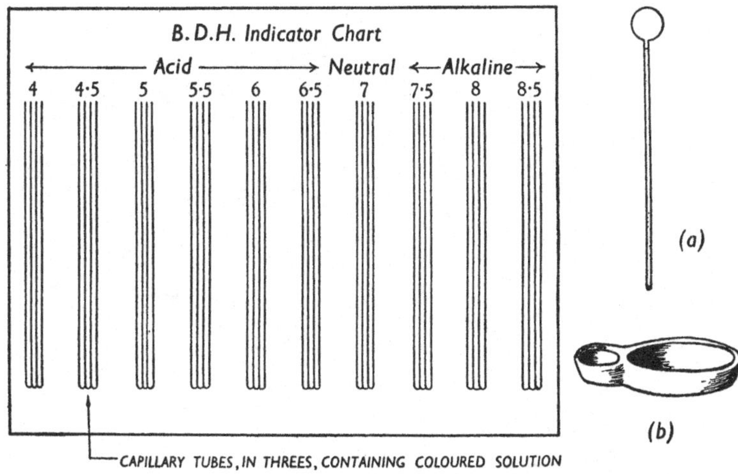

Fig. 74. *B.D.H. soil indicator chart for testing soil.* (*a*) *Capillary tube pipette;* (*b*) *porcelain boat*

less than 7 there is too much acid, if the number is above 7 there is too much lime.

(3) Lime also kills animals and plants (such as fungi) which harm crops.

Rotation of crops

Some plants use more of one kind of mineral, other plants use more of another kind. If the same crop is grown year after year in the same plot of land, the soil soon becomes deficient in the particular mineral required by the crop. To avoid this the farmer or the gardener ROTATES his crops, or sows different ones each year. Plants that have deeper roots may grow with plants that have shallow roots. It is difficult to clean the land of weeds if such crops as wheat are grown, but root crops, which have to be hoed several times, leave the land free from weeds.

CHAPTER 7

HOW A PLANT LIVES

Many processes are carried out by a plant in order that it may live. The study of these life processes is called PLANT PHYSIOLOGY.

There are four main activities that go on in a plant:

(1) The INTAKE of water and mineral salts by the roots of a plant.

(2) TRANSPIRATION, or the giving out of water by the plant, and the passage of water through the plant.

(3) PHOTOSYNTHESIS, or the building-up of food in the light with the storage of energy.

(4) RESPIRATION, which includes all the processes leading to the breaking down of food with the release of energy.

Intake of water

Experiment 18. Set up an experiment as shown in Fig. 75. Cover the water with a layer of oil so that no water can evaporate except through the plant. Mark the level of the water. You will notice that the amount of water decreases in the test-tube as the roots take in water. If a similar experiment is set up without the plant, the level of the water remains the same. The water is taken in by the ROOT HAIRS.

Many mineral salts are dissolved in small quantities in the water that is in the soil. Plants must take in an enormous amount of water to enable them to get the amount of mineral substances that they need. What happens to all the water that is taken in by the roots? Only a small amount is used by the plant, and the remainder passes through special conducting cells that are in the roots and stems to the leaves, where it escapes as water vapour through the PORES in the leaves.

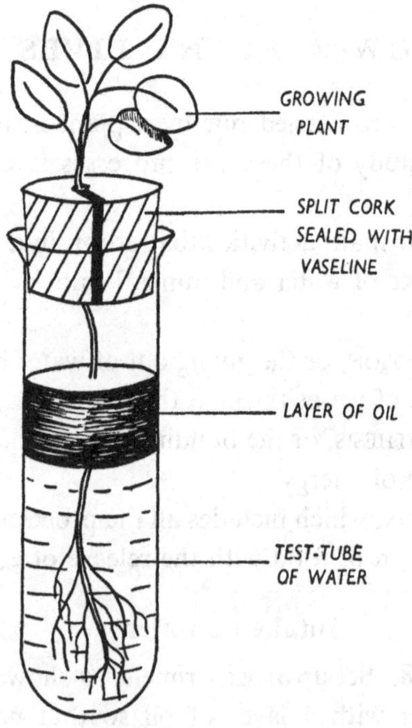

Fig. 75. *Experiment to show that roots take in water*

Experiment 19. Stand the stem of any small plant in a jar of red ink and leave it for a short time. Cut the stem across and downwards and you will see that the ink has risen only along certain channels where the conducting cells are. If you place the stalk of a white flower in red ink, you will see that the veins in the petals gradually become red.

Transpiration

There are several experiments that we may do which will show that leaves give out water.

Experiment 20. Take a potted plant that has been well watered and completely cover the pot and soil with a water-

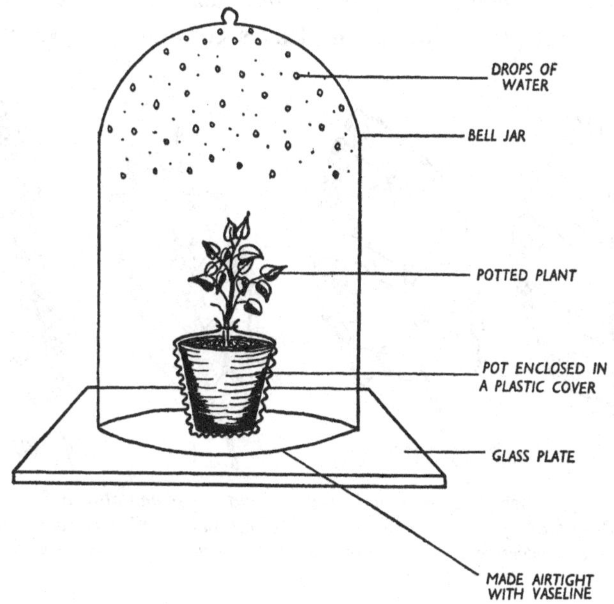

Fig. 76. *Experiment to show that leaves give out water*

proof bag to prevent water vapour from escaping from it. Stand it on a glass plate and place a bell jar over the whole thing, making the bottom of the bell jar airtight by smearing it with vaseline (Fig. 76). After a time you will see on the inside of the bell jar drops of water, which must have come out of the plant. Set up similar apparatus without the plant and no water will appear on the bell jar.

Repeat this experiment, but smother both sides of all the leaves with vaseline. You will see that no water is given out.

Water passes out of a leaf as water vapour. It can only pass through the pores or stomata as the surface of the leaf is covered with the cuticle which is formed of a waterproof substance.

If you pick leaves and do not put them into water they WILT, because the pores in the leaves continue to give out

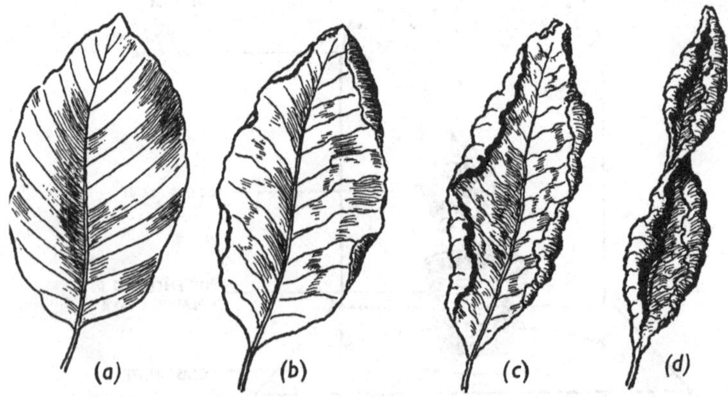

Fig. 77. *Experiment to show that leaves give out water and also to find out which side of the leaf gives out more water. (a) Both sides of a leaf smeared with vaseline; (b) underside smeared with vaseline; (c) upper side smeared with vaseline; (d) no vaseline. Hang up the leaves separately*

water which cannot be replaced. This can be shown in the following experiment.

Experiment 21. Take four similar leaves. Cover both sides of the first leaf with vaseline; cover the underside of the second leaf and the upper surface of the third leaf. Do not put any vaseline on the fourth leaf. Look at these leaves several hours later (Fig. 77). You will notice that the first leaf remains the same, as the pores are blocked, the second leaf withers very slightly, the third leaf withers fairly quickly, and the fourth leaf withers first as all its pores are open to allow water to escape. The third leaf withers before the second leaf, because there are more pores on the lower surface of the leaf than there are on the upper surface. Evergreen leaves have pores on the lower surface only.

A similar experiment may be done using cobalt chloride solution.

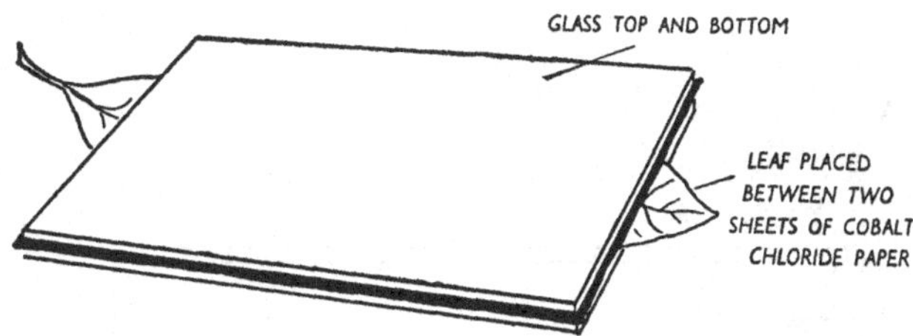

Fig. 78. *Experiment to find which side of leaf gives out more water*

Experiment 22. Soak some filter paper in cobalt chloride solution and dry it. The paper is pink when it is wet, and blue when it is dry. Take four leaves and vaseline them as in the previous experiment. Place each leaf between two pieces of dry cobalt chloride paper, and two pieces of glass. You must be quite sure that your paper is bright blue at the beginning of the experiment. Look at each side of the leaf, to see if the paper has turned pink. Compare your result with that of Experiment 21.

Experiment 23. Fig. 79 shows a form of POTOMETER that is used to measure the rate at which water is given off by leaves. The whole apparatus must be set up in a deep sink, under water, and the end of the leafy shoot must also be cut under water as there must not be any air bubbles. A bubble of air could be introduced into the bent tube by removing a drop of water with blotting paper. Take the apparatus out of the water and mark the position of the bubble. As the leafy shoot gives out water the bubble will move from *A* to *B*. The bubble can be sent back to *A* by opening the stopcock of the funnel. Take a piece of paper which is markėd off in inches and fasten it to the tube *AB*. By using a stop-watch you can find out how long it takes for the bubble to move 1 inch. Repeat this several times and take the average result. Place the

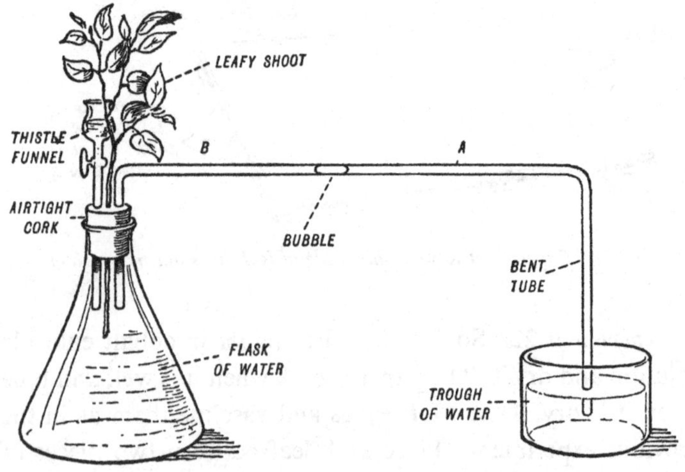

Fig. 79. *A potometer*

apparatus in a windy place, a sheltered place, in the sun and in the shade and find out whether the rate of transpiration varies under different conditions.

Prevention of loss of water by leaves

Leaves must not give out more water than the roots can take in. As autumn approaches, many trees, called DECIDUOUS plants (Fig. 80), lose their leaves and the breathing holes on the twigs close. By so doing they are getting rid of all the

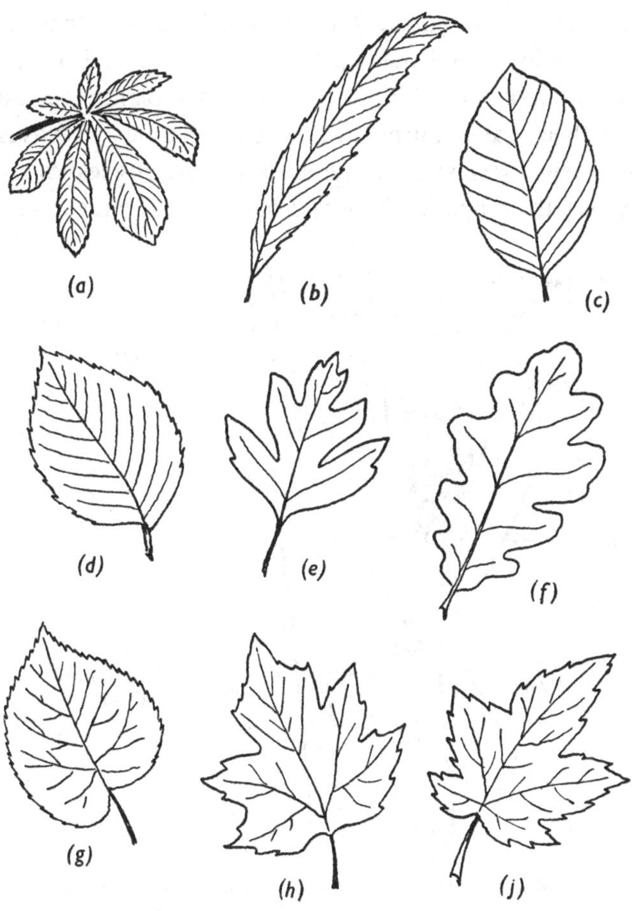

Fig. 80. *Leaves of some deciduous trees.* (*a*) *Horse chestnut;* (*b*) *sweet chestnut;* (*c*) *beech;* (*d*) *elm;* (*e*) *hawthorn;* (*f*) *oak;* (*g*) *lime;* (*h*) *plane;* (*j*) *sycamore*

pores through which water could escape. This is necessary because, during the winter when it is cold, the roots cannot take in as much water as they do during the spring and the summer. Before the leaf falls a layer of cork forms at the base of the leaf stalk to prevent the sap from coming out.

The walls of the cells outside the cork break down so that the cells become loose and the leaf falls off. This layer is called the ABSCISS LAYER (Fig. 81). EVERGREENS are plants that do not lose their leaves during the winter, because the leaves have very few pores on the lower surface only. Big leaves are usually very thick and leathery and have a thick cuticle, and

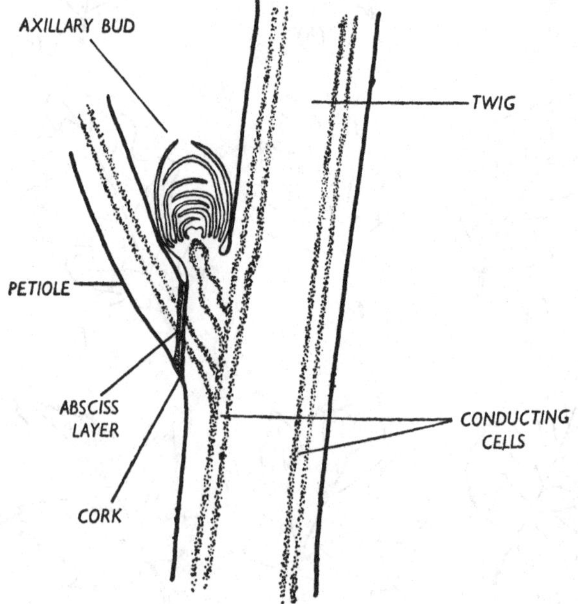

Fig. 81. *Diagram of a section through the base of a petiole just before leaf fall*

some leaves are very fine, so they can withstand the winter (Fig. 82). You could make a collection of the leaves of deciduous and of evergreen trees and bushes. You must press the leaves between blotting paper until they are flat and stiff. Dried leaves crack easily as they are very brittle, so you could make LEAF TRACINGS, which, when finished, look very real.

Place your leaf under a piece of paper, and scribble or rub over the leaf with a crayon or pencil that is the same colour as the leaf. Then cut out the leaf so drawn, and glue it on to your chart or into your scrapbook. A list of books is given in Appendix C which will enable you to identify your specimens.

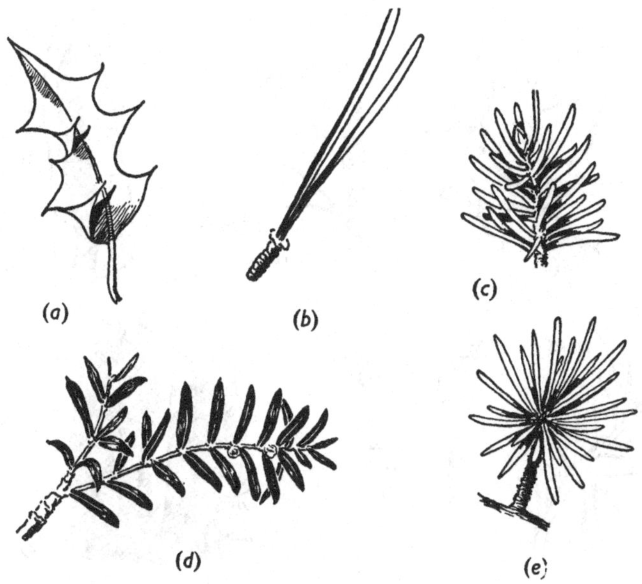

Fig. 82. *Leaves of some evergreen trees:* (*a*) *holly;* (*b*) *Scots pine;* (*c*) *spruce fir;* (*d*) *yew;* (*e*) *cedar*

The lower-growing or HERBACEOUS plants have various ways of preventing loss of water from the leaves during the winter. The shoots and leaves of many plants die, leaving only roots and buds in the ground, e.g. lupins, michaelmas daisies, phlox, chrysanthemums. Many plants die and leave only special storage organs in the ground. Some of these organs, such as dahlia tubers, potato tubers, crocus corms, iris rhizomes and onion bulbs were described in Book I. Some plants live for one year only, and are called ANNUALS. They die down in the autumn leaving only seeds in the ground. Plants such as daisies and grass that live close to the soil, where it is damp, do not change during the winter. Their leaves do not give out much water as they are living in a damp atmosphere.

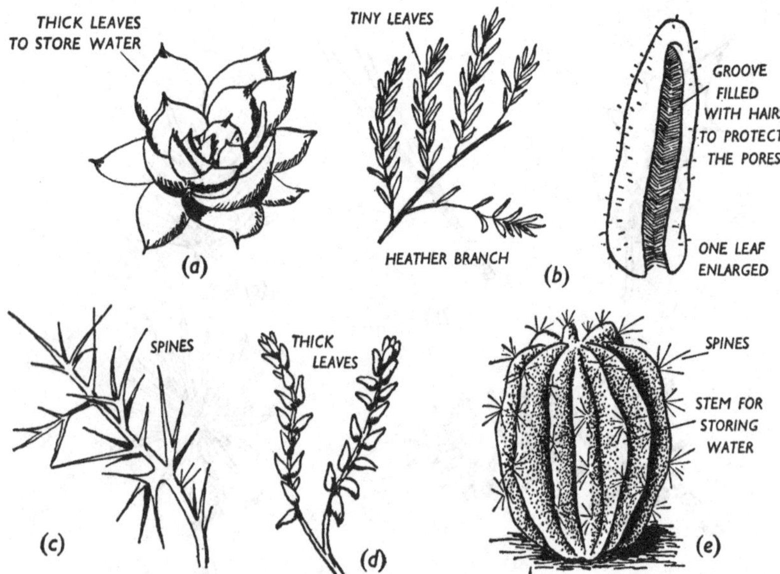

Fig. 83. *Leaves modified to prevent loss of water:* (*a*) *houseleek;* (*b*) *heather;* (*c*) *gorse;* (*d*) *stonecrop;* (*e*) *cactus*

There are many plants that live in very dry places, and they must not lose very much water at any time throughout the year. The leaves are specially modified in some way to prevent the loss of water. (1) The leaves may be very leathery, as are those of the evergreen plants. (2) The leaves may be densely covered with hairs to which the damp air clings, so that the leaves cannot lose water very quickly, e.g. houseleek, (Fig. 83*a*). (3) Heather has very tiny leaves, and the pores are hidden in a little hollow on the underside of the leaf (Fig. 83*b*). The hollow is covered with hairs. Many leaves have their pores hidden in pits or the leaves may be rolled to protect the pores. (4) The leaves may be reduced to tiny scales, e.g. broom, or to spines, e.g. gorse (Fig. 83*c*). (5) The houseleek, stonecrop and cacti are plants that store water in their cells during wet weather. Houseleek and stonecrop store water in their leaves, but cacti store water in their stems as the leaves are reduced to spines. This water can be used during a period of drought (Figs. 83*d* and *e*).

Fig. 84. *Hedges and trees shaped by the wind*

Have you ever seen trees similar to those shown in Fig. 84? The wind has made them grow like this. We all know that clothes dry quickly on a windy day, and similarly plants that grow in windy places dry up quickly and die. If new branches try to grow on the windward side, they lose too much water and die, so only the branches that grow on the leeward side of the tree can grow. This makes the tree one-sided.

Plants that live in salty places, such as salt-water marshes, are living in very dry places. Although there seems to be plenty of water, it is too salty for the roots to take it in. This can be shown by repeating Experiment 1 (page 127), using salt water in the test-tube.

Photosynthesis

Photosynthesis means the building-up or the making of food in the light. Plants can only make food in the LIGHT in those parts of the plant that contain CHLOROPHYLL. We know that the water that is taken in by the roots passes through the stems into the leaves. Carbon dioxide enters the leaves through the pores or stomata. In the cells that contain chlorophyll the water and carbon from the carbon dioxide join together to make sugar. This is changed to starch and is stored in the leaf during the day-time. When it is dark the starch, which will not dissolve in the sap, is changed into sugar, which will dissolve in the sap. It is then taken to all parts of the plant, where it may be stored as sugar or changed back again into starch. This process of making food is very important to all animals and plants because, from these simple foods, more complicated foods are made which they need. Living things need energy to carry out the processes that are necessary for them to live. All energy comes from the sun. During photosynthesis, plants store energy in the food which they make.

The easiest way to show that photosynthesis is taking place is to test for starch in a leaf.

Experiment 24. To test for starch. Pour very dilute iodine on to starch and you will see that it turns a bluish-black colour. Then pour iodine on to any substance that does not contain starch and you will see that it remains brown.

Before we can test for starch in a leaf we have to remove the chlorophyll.

Experiment 25. To remove the chlorophyll. First boil the leaves to kill them, and place them in a test-tube containing ethyl alcohol. Stand this tube in a beaker of water that has just been boiled. (Do not leave the burner under the beaker.) This will remove the chlorophyll. Wash the leaves and put them in a dish containing iodine. Rinse the leaf. It will be black if it contains starch.

The following experiments will show you that photosynthesis will only take place in the LIGHT, in the presence of CHLOROPHYLL, and that CARBON DIOXIDE and WATER are necessary for starch to be made.

Experiment 26. To show that chlorophyll is necessary. Take a variegated leaf, that is, one that is mottled green and white. A deciduous leaf is better than an evergreen one. Draw the leaf, marking the yellow and the green patches. Then remove the chlorophyll and test the leaf for starch. Compare the black and brown patches on the leaf with the original drawing and you will find that the black marks denoting the presence of starch are seen where the chlorophyll was present in the leaf.

Experiment 27. To show that carbon dioxide is necessary for photosynthesis. Place a shoot in the dark for 24 hours to remove the starch and then set up the apparatus as shown in Fig. 85, leaving some leaves of the shoot inside the jar, and some leaves outside the jar. The small test-tube of caustic potash is put into the jar because it absorbs all the carbon dioxide from the air that is inside the jar. The soda lime removes the carbon dioxide from the air entering the jar. Leave the apparatus for several hours in the light and then test the leaves that were inside the jar, and those that were outside the jar, for starch. You will see that the leaves inside the jar which had not any carbon dioxide do not contain starch.

Experiment 28. To show that light is necessary for photosynthesis. Take a potted plant and place it in the dark for 24 hours. Cover one leaf with silver paper and then leave the plant in the light for several hours. Test the leaf that was

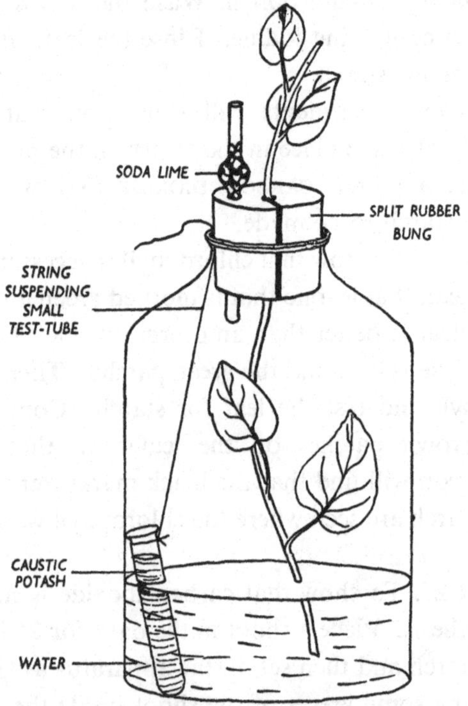

Fig. 85. *Experiment to show that carbon dioxide is necessary for photosynthesis*

covered and one other leaf for starch. You will see that the leaf that had not any light does not contain starch.

All the leaves on a plant are so arranged that they do not overlap, so each leaf gets as much light as possible.

Experiment 29. To show that water is necessary for photosynthesis. Put a potted plant, whose leaves have wilted through lack of water, in the dark for 24 hours to remove all the starch. Then place it in the light for several hours. Test the leaves for starch and you will find that they do not contain any starch.

During photosynthesis, when carbon dioxide and water join together to form sugar, there is a surplus of oxygen which

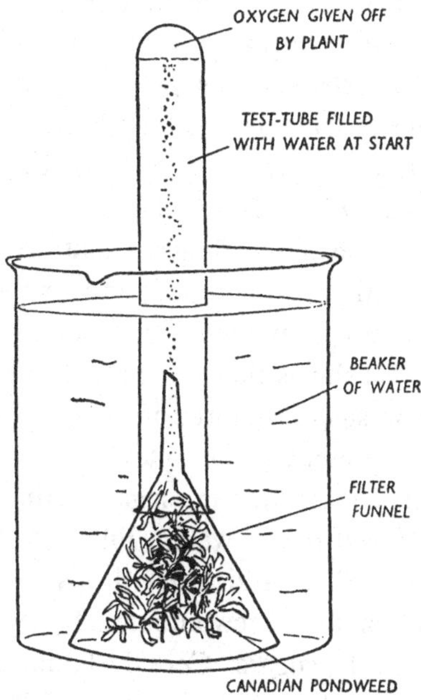

Fig. 86. *Experiment to show that leaves give out oxygen*

is givėn out through the stomata. The following experiment will show you that oxygen is given out during photosynthesis.

Experiment 30. Set up the apparatus as shown in Fig. 86, using a water plant such as Canadian pondweed. Set up the apparatus under water to ensure that there are no bubbles in the test-tube. Then place the apparatus in the light and leave it for several days. Bubbles of gas will rise out of the leaves and will collect in the test-tube. When the test-tube is full of gas remove it from the apparatus and place a glowing splint into it. The splint immediately re-lights, showing that the gas is oxygen.

Tests for foods stored in plants

We have learnt that sugar is made in the leaves of plants. This may be stored as sugar or as starch, or from it more complicated substances such as proteins and fats may be formed. Experiment 24 shows us how to test for starch.

Experiment 31. To test for glucose. Grind up the plant to be tested, in water. Filter off the liquid and place in a test-tube. Add a few drops of Fehling's A and then a few drops of Fehling's B solution to the liquid, and gently heat it. The blue colour gradually disappears and an orange-red precipitate is formed. This shows that glucose is present.

Sometimes cane sugar is stored in plants. You can show that cane sugar is present in the following way.

Experiment 32. To test for cane sugar. Grind up the plant to be tested in a few drops of water. Filter off the liquid and pour it into a test-tube. Add a few drops of dilute hydrochloric acid and boil it for a few minutes. This will change the cane sugar into simpler sugars. Cool the liquid and neutralize the acid by adding sodium bicarbonate until it stops fizzing. Then test for glucose as in Experiment 14.

Experiment 33. To test for proteins. Grind up the plant to be tested in water and filter off the liquid. Pour it into a test-tube and add a few drops of Millon's reagent. A precipitate is formed which turns pink when it is heated.

You can also test for proteins by adding a few drops of caustic soda and one drop of copper sulphate solution. A violet colour appears.

Experiment 34. To test for fat. Take small pieces of the plant to be tested and press them on filter paper. Dry the filter paper. If fat is present there will be a greasy mark.

Osmic acid, which is colourless, will turn black if fat is present.

Respiration

All living things RESPIRE or breathe. They do this so that they can obtain energy to carry out all the functions that are necessary for them to live. Most of you know that we need energy to move, to run, etc., but you may not realize that energy is needed to breathe, to eat, to think and indeed for all the functions that go on in a living thing. Plants also need energy. All living things derive their energy from the sun. We have already learned that some energy from the sun is stored in the food during photosynthesis. This energy is released when the foods that have been made are broken down again either in the plant, or in an animal which has eaten a plant. Oxygen is necessary for the breaking down of food and for the release of energy. When food is broken down, water and carbon dioxide are formed. The carbon dioxide passes out through the stomata.

Experiment 35. To show that plants respire. Place some flowers in a flask and invert it over a dish of mercury. By means of a pipette, put caustic potash into the flask, as shown in Fig. 87. Mark the position of the top of the mercury. As the oxygen in the flask is used up, carbon dioxide is given out which is absorbed by the caustic potash. The mercury rises up the flask to replace the gas that has been used.

Experiment 36. Put some lime water into a test-tube and blow down a piece of glass tubing into the lime water. The carbon dioxide which you breathe out turns the lime water milky.

In the following experiment you can show that peas are respiring by showing that they are giving out carbon dioxide.

Experiment 37. To show that soaked peas give out carbon dioxide during respiration. Set up the apparatus as shown in Fig. 88, making sure that all the connexions are airtight. Fix a filter pump or an aspirator full of water (see Fig. 67) to the end indicated to draw air through the apparatus. The soda lime and caustic potash remove the carbon dioxide from the air entering the apparatus. The lime water in flask 2 remains clear, showing that there is no carbon dioxide entering flask 3 which contains the peas. The lime water in flask 4 turns milky showing that the peas have given out carbon dioxide.

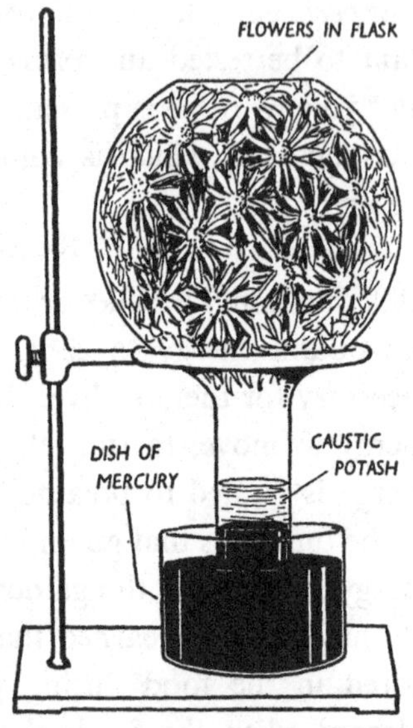

Fig. 87. *Experiment to show that plants respire*

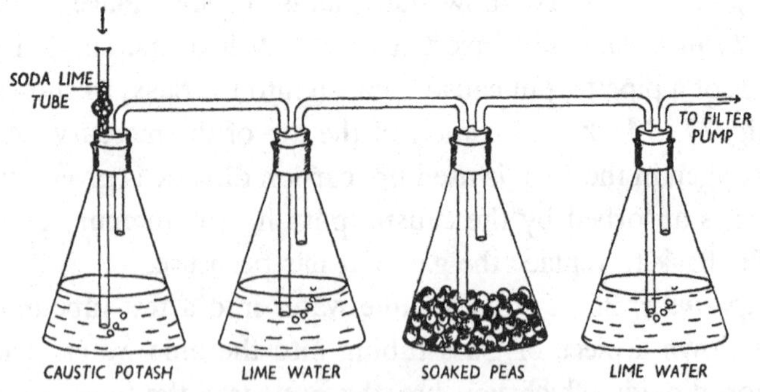

Fig. 88. *Experiment to show that peas give out carbon dioxide during respiration*

CHAPTER 8

HUMAN PHYSIOLOGY

In this book we cannot describe the structure of all the animals that we have seen, or explain how the various organs function. Instead, we shall study our own bodies and learn how they work.

The skeleton and its functions

All vertebrate animals have bones which support the body. In our bodies we have about two hundred bones which not only support the body, but which also protect the important parts of the body, and act as rigid levers to which muscles are attached.

If you feel your SKULL you will probably think that it is made of one piece of bone. Try to get the skull of an animal and look at it closely. You will see that it is really made up of a number of bones which are joined together by saw-like edges. The bones in a baby's skull are not joined together; this enables the head to grow. The skull protects the BRAIN. The lower jaw is joined to the skull in such a way that it moves upwards and downwards very easily. The TEETH (see page 151) grow in the upper and in the lower jaws. A thick bundle of nerves called the SPINAL CORD passes out of the brain through a large hole in the base of the skull. The spinal cord is protected by the SPINAL COLUMN, which consists of thirty-three small bones called VERTEBRAE. The first twenty-

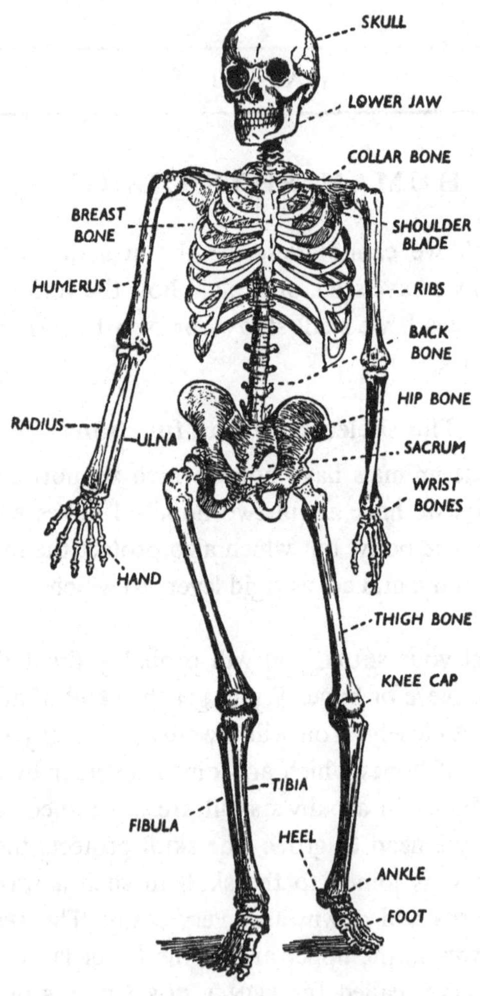

Fig. 89. *The human skeleton*

four are joined by rings of cartilage which enable us to bend our backs, but the last nine vertebrae are joined together to form a solid piece of bone, the SACRUM.

Each vertebra consists of a solid knob of bone, the CENTRUM, which faces inwards to the body, and a bony arch on top of it, the NEURAL ARCH, which faces outwards (Fig. 90). This ring of bone surrounds a space, the NEURAL CANAL through which the spinal cord runs. Three bony projections grow out of the neural arch: the NEURAL SPINE in the middle, which you can

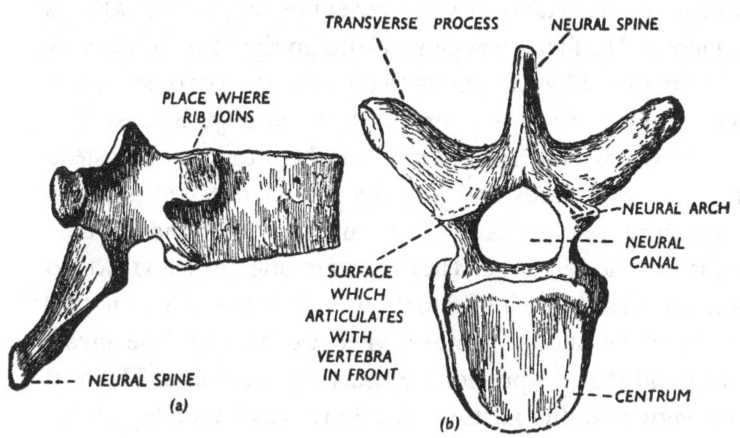

Fig. 90. *A vertebra: (a) side view, right side; (b) view from above*

feel through your flesh and two lateral ones, the TRANSVERSE PROCESSES. Muscles are joined to these bony projections. The vertebrae are joined together by discs of cartilage which make the back flexible. Each vertebra has two surfaces at the front which point upwards and which articulate with two similar surfaces which point downwards on the vertebra in front of it. These surfaces, together with ligaments, keep the vertebrae in position. There are holes at the side, between the vertebrae, through which nerves from the spinal cord pass. On either side of the hole in the skull there is a large round knob which fits into the hollows of the first vertebra, which is called the ATLAS. This enables us to nod our heads. The second vertebra, which is called the AXIS, fits into the first in such a way that we can move our heads sideways. There are seven vertebrae in the neck including the atlas and axis, which are called the CERVICAL vertebrae. They can be recognized by a tiny hole on either side of the neural canal, through which a small blood vessel passes. To each of the next twelve vertebrae, which are called THORACIC vertebrae, a pair of RIBS is attached. The first seven pairs of ribs go right round the body and are joined by CARTILAGE to the BREAST BONE or STERNUM. You can feel the breast bone in the front of your chest. The next three pairs of ribs do not join the breast bone; instead they curve upwards and join the pair of ribs that is above them. The last two pairs of ribs are very short and are only joined to the vertebrae. These ribs are often called FLOATING RIBS. As the ribs are joined to the breast bone or to one another by cartilage, they can move when we breathe. The breast bone and the ribs protect the heart and the lungs. The next five vertebrae, the LUMBAR vertebrae, have very big centra and extra bony projections for the attachment of muscles, as they have to support the heavy abdominal part of the body, and the legs. The next five vertebrae are joined together to form the SACRUM, and the last four are also joined to the sacrum and bend inwards to form the 'tail' or COCCYX.

There are two girdles of bones which support the limbs. The SHOULDER or PECTORAL girdle supports the arms. It is made up of two COLLAR bones or CLAVICLES at the front and two SHOULDER BLADES or SCAPULAE at the back. You can easily feel your two thin collar bones at the base of your neck. Each one is joined at the front to the breast bone and at the other end to the shoulder blade. The shoulder blades are triangular in shape, with the base of the triangle facing towards the backbone. On the outer surface of each one there is a ridge to which muscles are attached. They are not joined together; if they were you would not be able to bend your shoulders forwards. At the corner of the triangle there is a hollow into which fits the rounded end of the bone of the upper arm, which is called the HUMERUS. At the elbow the two bones of the lower arm, which are called the RADIUS and ULNA, move on the end of the humerus. An extension of the ulna forms the elbow and prevents the lower arm from moving backwards. The radius is on the thumb side of the arm. When you turn over your hand the radius moves over the ulna at the wrist end (see Fig. 89). There are eight bones in the WRIST called CARPALS and five in the PALM of the hand called METACARPALS. Each FINGER has three bones and the THUMB has two bones.

The HIP GIRDLE or PELVIS consists of a number of bones which are fused together and join the sacrum of the backbone. The pelvis, which is wider in a woman than it is in a man, has to support the weight of the body. On either side of the pelvis there is a socket into which the rounded head of the thigh bone or FEMUR fits. The legs are similar in structure to the arms. There is one bone in the upper part of the leg—the femur—and two in the lower part, which are called the TIBIA and FIBULA. The fibula is on the same side of the leg as the big toe. There are seven bones in the ANKLE called the TARSALS, one of which forms the HEEL. There are five bones in the INSTEP called the METATARSALS and three bones in each of the LITTLE TOES, and two bones in the BIG TOE. There is a bone in the knee which is called the KNEE CAP or PATELLA. This bone prevents the leg from bending backwards as well as forwards.

The bones of a baby and of a young child are made of cartilage and can easily be bent. As the child gets older the cartilage is gradually changed into bone.

Muscles

We cannot see muscles, but we can feel some of them working as we move our bodies. Muscles are joined to bones by TENDONS or SINEWS. They can expand and contract,

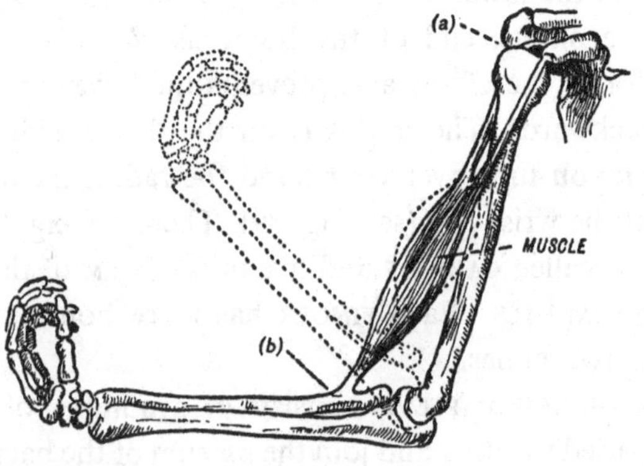

Fig. 91. *Action of the biceps muscle. The muscle is attached by two tendons to the shoulder at (a), and by one tendon to the forearm at (b). When the muscle contracts the arm is drawn up*

and by so doing they may move other parts of the body. Stretch out your right arm and hold the muscular part of your upper arm with your left hand. Then slowly raise your right hand. You will feel the muscle, which is called the BICEPS, getting fatter and shorter. As you lower your hand the muscle will expand again (Fig. 91). The muscles of the HEART and of the INTESTINE are joined to other muscles to form a hollow tube. As the muscles contract the tube becomes narrower and blood or food is forced along the tube. As the muscles relax, more blood or food can enter. These muscles, which move on their own, are called INVOLUNTARY muscles. All the muscles in the body which we can move when we so desire are called VOLUNTARY muscles. When we are born we have no control over our muscles, but we gradually gain control over many of them.

Teeth

When you have all your teeth you will have thirty-two. Count the number of teeth that you have now. The teeth on

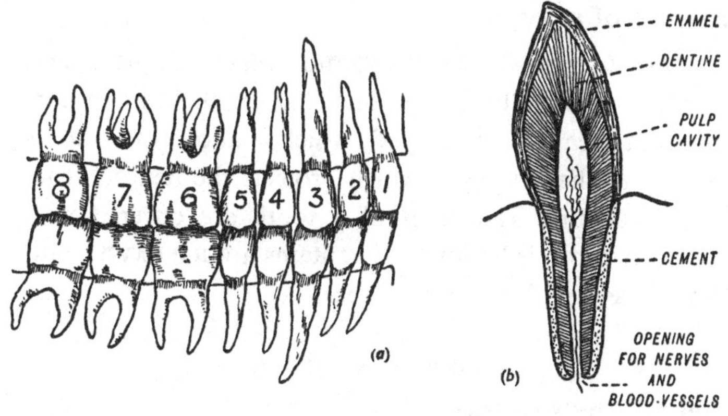

Fig. 92. *Human teeth.* (*a*) *The right side of the upper and lower jaw, showing the teeth:* 1 *and* 2 *are incisors;* 3 *canine teeth;* 4 *and* 5 *pre-molars;* 6, 7 *and* 8 *molars;* (*b*) *section through a tooth*

each side of the top and of the lower jaw are similar, so we will describe the eight on one side (Fig. 92). The two front teeth (or INCISORS) have a sharp edge for biting food. Each one has a single FANG which holds the tooth in the jaw. The next tooth (the CANINE tooth) is similar to an incisor, but it is much bigger. The next five teeth are 'double' teeth. The first two are PRE-MOLARS and each one has two fangs, the last three are MOLARS and they have three or four fangs each.

Babies usually have no teeth when they are born. From six months onwards the teeth gradually appear until there are twenty in all. These teeth are called the MILK TEETH. Young children have not any molars. The milk teeth become loose at any time after a child is six years old and they are replaced by the PERMANENT teeth. In addition to the twenty which replace the milk teeth, two molars appear on each side, making twenty-eight teeth. Later on the last molars or WISDOM teeth appear.

Structure of a tooth

A tooth consists chiefly of DENTINE, which is a hard substance that is something like bone (Fig. 92*b*). The top part of the tooth (called the CROWN), which is above the gum, is covered with hard ENAMEL which protects the dentine from harmful bacteria. The fangs of a tooth are covered with CEMENT. A tooth is hollow and contains a PULP CAVITY into which nerves and blood vessels pass.

The digestive system

Different kinds of food

If we wish to remain healthy we must be sure that we are eating the right kinds of food. Our diet should include CARBOHYDRATES, PROTEINS, FATS, MINERALS and VITAMINS.

Carbohydrates. Sugars and starches belong to this group. It is from these foods that we get our ENERGY (see page 138). All foods pass eventually into the blood after they have been digested. The blood carries the food to all parts of the body. If there is more sugar in the blood than the body requires it is taken out of the blood in the LIVER. It is changed to a special kind of animal starch called GLYCOGEN and is stored in the liver. At any time this glycogen can be changed back again into sugar and passed into the blood, to be sent to those parts of the body which need it. The liver, then, keeps a constant control on the amount of sugar that is in the blood. It can do this only if it receives a supply of a substance called insulin, which is produced in the pancreas (see page 168) and carried to the liver by the blood. Without insulin a person suffers from sugar diabetes.

If there is too much glycogen for the liver to store it is changed into fat and is stored elsewhere in the body.

Fats. Fatty foods give us WARMTH. We also get some energy from fats, but not as much as we do from carbohydrates. Fat may also be used to help to build new cells in the body, but it is chiefly RESERVE FOOD that is stored below the skin, between the muscles, and round the internal organs. If a person fasts, the body gets thinner because the store of fat is used up. The fat in the skin helps to insulate the body so that the body does not lose heat. Whales are warm-blooded mammals, and have much fat or blubber in their skins to keep them warm.

Proteins. These give us a certain amount of heat and of energy, but their chief value is in BUILDING UP new parts of the body or in replacing worn tissues. We get our protein food from milk, eggs, lean meat, fish, peas, beans, gelatine, etc. Surplus protein cannot be stored.

Minerals. Many mineral salts are necessary for the good health of the body. CALCIUM, which is found in milk and cheese, hardens the bones and the teeth and prevents rickets. It also helps in the clotting of the blood and in the functioning of the nerves. PHOSPHORUS, which we get from cheese, milk, wholemeal, oatmeal, barley, eggs, meat, fish, pulses and nuts, also helps to harden the bones and the teeth, and it is necessary to make good blood and healthy nerves. IRON, which we obtain from wholemeal, oatmeal, eggs, potatoes, green vegetables (especially watercress) and red meat, is necessary for the formation of HAEMOGLOBIN, which is the red substance in the blood. It is the haemoglobin that carries the oxygen round the body. IODINE is necessary for the proper working of the thyroid gland (see page 168). It is present in fish, and is today present in iodized salt. MANGANESE stimulates the body to produce its own antidotes against disease. There are many more minerals in the body, but we know very little about most of them. Minerals are often added to foods that we buy to ensure that we eat sufficient of them. Some salt that you buy contains a little calcium, phosphorus and magnesium carbonate. Look on the labels of any food that is given to children or invalids and you will see that many of them contain these important minerals.

Vitamins. Until comparatively recent times it was thought that carbohydrates, fats, proteins and minerals were the only foods necessary to keep the body healthy. Animals were given pure extracts of these foods and it was found that the animals were not well, because they lacked certain substances which were called vitamins. Many different vitamins have been found, but only the chief ones are mentioned in this book. Today many vitamins as well as minerals are added to our foods, and particularly to the foods of babies and mothers. The chief vitamins are A, B, C and D. All these vitamins are found in the plant kingdom, so that fruit and vegetables are good sources of vitamins.

Water is also very essential.

The principal vitamins

Vitamin A. This vitamin is important because it affects the rate of growth of the body and also the resistance to disease.

It is also necessary to make the eyes see at night. Without it people suffer from night-blindness. Vitamin A is added to many foods, especially baby foods. It is present in milk, cheese, cream, egg yolk and carrots.

Vitamin B. At one time it was thought that this was a single substance, but it is now known to consist of a number of factors which are found in similar types of food, such as whole cereals, dairy produce, yeast and vegetables. The chief factors are vitamin B_1, vitamin B_2, vitamin B_3 and vitamin B_{12}.

VITAMIN B_1. The lack of this vitamin causes loss of appetite, constipation and pains in the stomach, a weaker heart-beat, and low blood pressure. A person may have palpitation and be breathless, there might be neuralgia, neuritis or even complete paralysis of the nerves. We can tell if a person is having enough of this vitamin by testing the urine.

VITAMIN B_2 is essential for growth.

VITAMIN B_3. Lack of this vitamin causes dermatitis on the face and hands and any part of the body that is exposed to the light. It also causes inflammation of the digestive tract and diarrhoea. It may cause a disturbance of the brain and of the nerves.

VITAMIN B_{12}. Lack of this vitamin causes anaemia as it is essential for the proper development of the red corpuscles.

Vitamin C. This gives us resistance to disease. Lack of it will cause haemorrhage, bleeding gums, painful joints and fragile bones. It is present in green vegetables, salads, dairy produce, tomatoes, blackcurrants and oranges.

Vitamin D. It enables the bones and the teeth to develop properly. Without it children will have rickets. The bones become soft, fragile and deformed. Vitamin D prevents the decay of the teeth. Our main source of vitamin D is foods such as egg yolk, cod liver oil and salmon, but in addition we can obtain a small quantity from a substance in our skin which can be changed to vitamin D by the sun. So it is important that we should get as much fresh air and sunlight to our bodies as we can without burning our skins.

Vitamin E. It is present in lettuce and whole wheat and is necessary for the production of young.

Vitamin K. It is present in liver, cabbage, spinach and tomatoes, and helps the blood to clot.

Testing for foods

By carrying out the tests given on pages 142 and 143 you can easily find out whether certain types of food contain starch, sugars, proteins or fats. It is not easy to test for minerals or for vitamins, so we shall not attempt these experiments here.

Digestion

Have you ever wondered what happens to your food after you have put it into your mouth? It has to be digested, that is, changed into a soluble form so that it can pass into the blood, which carries it to all parts of the body. The food is digested by substances called ENZYMES which are produced in different parts of the digestive organs. The food passes along a tube called the ALIMENTARY CANAL, whose walls are very muscular. The muscles alternately expand and contract to push the food along the tube. The digestive organs secrete a substance called MUCIN which lubricates the food.

In the mouth the food is ground up into small pieces by the double teeth, so that it will be more easily digested. It is also mixed with SALIVA, a juice which is produced in six glands and passed down small tubes into the mouth. Four of these glands are under the tongue and the other two (which swell when you have Mumps) are between the angles of the jaw, where the jaw joins the skull. The saliva contains an enzyme called PTYALIN which begins to digest starch. The food is then swallowed, and a flap of skin (which is called the EPIGLOTTIS) closes over the tube that leads to the lungs and so prevents the food from going the 'wrong way'. The food passes down a tube called the OESOPHAGUS into the stomach. The walls of the stomach produce a liquid called the GASTRIC JUICE which contains hydrochloric acid and enzymes. The acid stops the action of ptyalin, kills any bacteria that may be present in the food, breaks down cane sugar to simple sugars which can dissolve, and also helps the enzymes to work. The enzymes are PEPSIN which begins the digestion of proteins, and RENNIN which coagulates milk. By muscular action of the stomach the food is churned up. At the far end of the stomach there is a very tight muscle which prevents the food from passing on until it is nearly digested. Some foods (for instance, fats) take longer to pass through the stomach than do other foods. As the food is digested the muscle relaxes and lets some food through to the DUODENUM. In the duodenum there are two juices which continue to digest the food. (1) The BILE, which is made in the LIVER and is stored in the GALL BLADDER. As food passes out of the stomach some bile is sent down the bile duct into the duodenum, where it helps to digest fats. Bile contains a substance called sodium bicarbonate, which neutralizes the acid from the stomach, as all the remaining enzymes cannot work if acid is present. (2) A second juice (called the PANCREATIC JUICE) is made by the PANCREAS, which lies along the base of the stomach (Fig. 93). This juice contains enzymes which digest all kinds of food. (Carbohydrates are digested by amylopsin, fats by steapsin, and proteins by a substance which is changed to the enzyme trypsin in the ileum.) In Appendix B a number of experiments are given which will show the action of some enzymes on your food.

The food then passes into the next part of the SMALL INTESTINE called the ILEUM, which is a coiled tube in the abdomen. It is rather narrow, but it is 22 feet long. In the ileum digestion continues as the walls secrete several enzymes

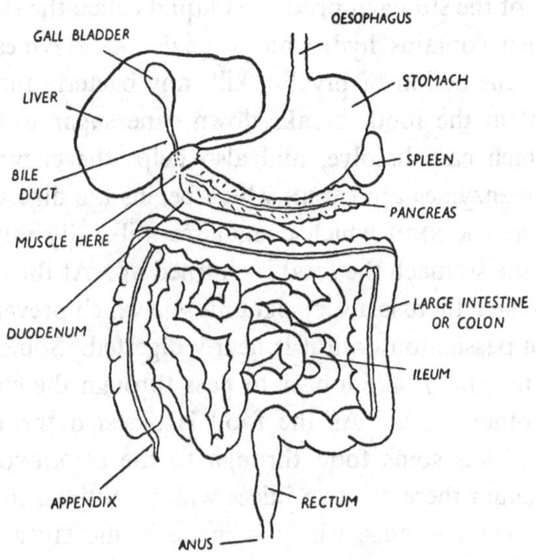

Fig. 93. *Diagram of the alimentary canal*

which digest all kinds of food. (Maltase completes the digestion of starch, erepsin digests proteins and invertase turns cane sugar into simple sugar.)

The inner wall of the ileum has transverse folds to increase the area of the surface. Each fold has numerous finger-like projections called VILLI, which absorb the digested food. Each villus has a central vessel called a LACTEAL which is surrounded by blood capillaries. Digested carbohydrates and proteins pass into the blood, but the digested fats pass into the lacteal which contains a liquid called LYMPH (see page 164). The lacteals unite to form larger vessels which join and eventually enter a blood vessel on the left side near to the heart.

The small intestine then enters the LARGE INTESTINE OR COLON. This is a shorter but fatter tube. The APPENDIX grows out of the first part of the colon. In the large intestine water out of the undigested food is absorbed. The remainder of the solid waste food goes into the RECTUM and is passed out through the ANUS, at least once a day, as FAECES. If the faeces are not got rid of a person becomes constipated. Then poisonous substances pass from the waste food into the blood and the person becomes ill.

Respiration

You must always breathe through your nose and not through your mouth. Hairs in the nostrils catch the dust particles which contain the germs, and so prevent them from entering the body. These hairs must be kept clean to do their job properly. As the air passes through the nose it is warmed by the many blood vessels that are there. The air passes down the THROAT into the WINDPIPE OR TRACHEA, which divides into two BRONCHI (Fig. 94). Each bronchus leads into a LUNG. The trachea and the bronchi are strengthened by incomplete rings of cartilage, which you can feel in the front of your neck. The ring of cartilage is broken at the back to allow the food to pass down the oesophagus. In each lung the bronchus branches into BRONCHIOLES which divide again until the tubes become very fine. Air sacs called ALVEOLI open from these fine tubes, so an enormous area of lung surface is in contact with the air (Fig. 95). This lung surface is covered with a network of fine blood vessels or CAPILLARIES into which the oxygen passes from the air. Oxygen is carried by the red corpuscles in the blood to all parts of the body, where it passes out of the blood into the cells in solution in the plasma which is the liquid part of the blood. In the cells, the

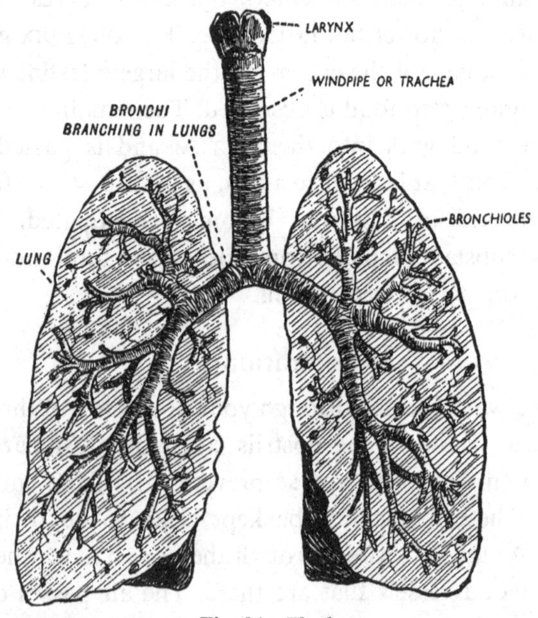

Fig. 94. *The lungs*

oxygen is used to break down food so that energy can be released. Carbon dioxide is given off which is carried by the plasma of the blood to the lungs, where it passes from the capillaries into the air sacs and is then breathed out. The lungs are in the THORAX and, as you know, they are protected by the ribs, the backbone and the sternum. A muscular DIAPHRAGM separates the thorax from the abdomen. The ribs are movable because they are joined to the breast bone by cartilage. When we are about to breathe in, muscles move the ribs upwards and forwards and the diaphragm moves downwards. This makes the thorax bigger and so the air rushes into the lungs. When the muscles relax the ribs and the breast bone fall, the diaphragm arches upwards, and so air is forced out of the lungs.

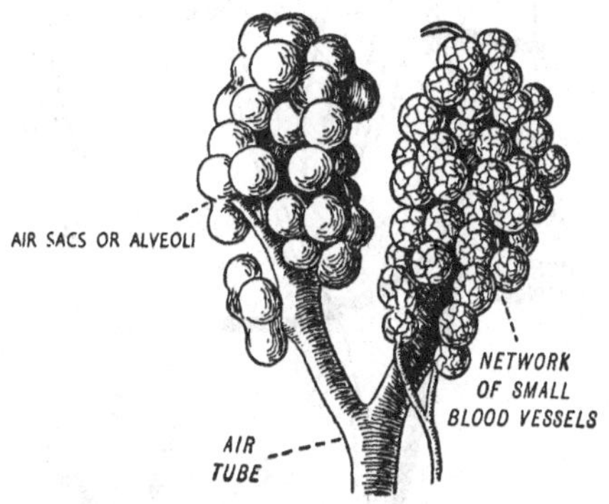

Fig. 95. *Diagram showing air sacs in the lungs and arrangement of blood vessels round the air sacs*

The blood system

The blood system is the transport system of the body.

Blood flows throughout the body in blood vessels. The HEART (which is the muscular, pumping organ) pumps the blood round the body. It lies in the THORAX between and in front of the lungs. The lower, pointed end of the heart is pointed to the left and is not central. The heart is divided longitudinally into two completely separate halves. Through the LEFT half of the heart passes the blood which has come from the lungs and which contains oxygen. This blood is bright red in colour. The RIGHT half of the heart receives blood from all parts of the body. This blood contains carbon dioxide and is a dull bluish-red in colour. Each half of the

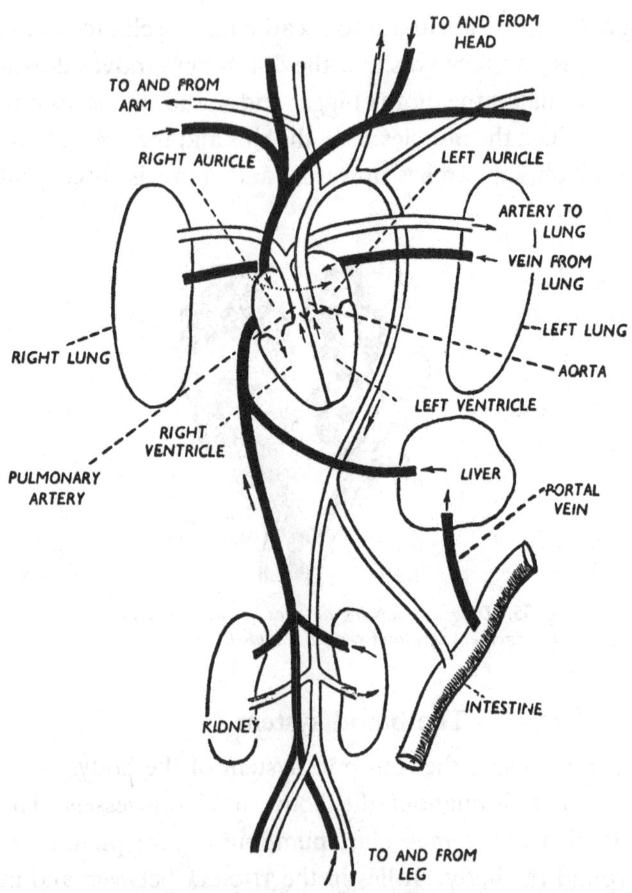

Fig. 96. *Diagram of human heart and blood vessels, seen from the front. Veins are shown black, arteries white. The valves at the openings into and out of the ventricles are indicated. The course of the circulation is shown by arrows*

heart is divided into a top half, the AURICLE, and a lower part, the VENTRICLE. The auricles are thin-walled and receive blood from the veins. The ventricles have thick muscular walls which contract to pump the blood into the arteries. The wall of the left ventricle is very thick since it has to pump blood to all parts of the body. There are flaps of tissue forming valves between the auricles and the ventricles, which prevent blood from returning to the auricles when the ventricles contract. The valve on the left side of the heart has two flaps and is called the BICUSPID valve, the one on the right has three flaps and is called the TRICUSPID valve. Both valves are held down by cords which are joined to the walls of the ventricles.

Let us follow the course of the blood round the body after it has passed into the left auricle from the lungs. As the auricle fills with blood the muscular walls stretch as far as they can, then they contract and force the blood into the ventricle. When this contracts blood is forced into the AORTA which gives off branches, the thick-walled ARTERIES, which take blood to all parts of the body. A valve with three pockets (called the SEMILUNAR VALVE) prevents blood from passing back into the ventricle. The rhythmical expansion and contraction of the muscles of the heart gives the HEART BEAT. Where the arteries come to the surface of the body (as they do in the wrists and the temples, and many more places) you can feel the pulse. It beats at various rates depending on your age, how active you are, and various other circumstances. At 14 years of age the average is 80 to 85 and for adults 70 to 80. During the first year of its life the beat of a baby averages 120 beats per minute. The arteries get narrower as they get farther away from the heart, and end as very fine CAPILLARIES. (If you roll a live frog in a piece of paper to prevent him from wriggling, and spread out the web of a hind foot under a microscope, you will be able to see the capillaries with the blood moving along. This will not hurt the frog if you handle it carefully.) The blood is brought back to the right auricle in the veins, which are thin-walled. The force of the heart beat is lost in the capillaries, so the veins have valves which prevent the blood from flowing in the wrong direction (Fig. 97). The muscular action of the body helps to send the blood along the veins. All the veins join to form two large veins which enter the right auricle. Blood passes from the right auricle into the right ventricle which pumps the blood (along the PULMONARY ARTERIES) to the lungs.

Some of the liquid in the blood oozes through the capillary walls and surrounds the neighbouring cells. This liquid can get back to the heart in two ways: (1) it can pass back into the blood capillaries; (2) it may find its way into special LYMPH capillaries, which are not connected with the arteries but have blind ends. They gradually join together to form bigger lymph vessels which, like the veins, have very thin walls and also possess valves. At frequent intervals along the lymph vessels there are special groups of cells which are called LYMPH NODES. Nearly all the lymph vessels of the body finally join together to form one large lymph duct, which enters a large vein in the region of the left shoulder. A small lymph duct on the right side carries lymph from the right arm and the right side of the head and thorax. Muscular action of the body forces the lymph along the vessels.

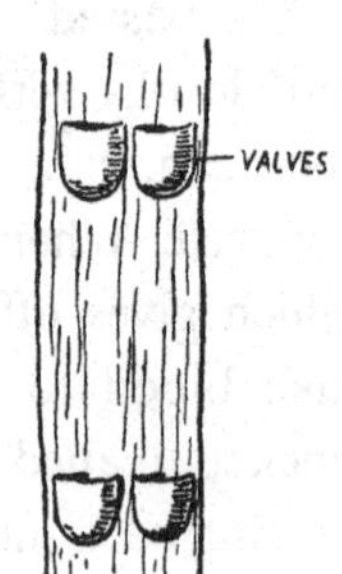

Fig. 97. *Valves in veins*

Composition of the blood

Blood consists of a colourless liquid called PLASMA, in which float a number of cells called CORPUSCLES. Some corpuscles are RED and are rather like pennies in shape (Fig. 98*a*). They are formed in the red BONE MARROW, which is the soft substance found inside some bones. It was thought that these cells lived for about twenty-five days, but recent experiments seem to show that they live from seven to ten weeks, and are then destroyed by special cells which are in the blood vessels in the liver and in the SPLEEN (Fig. 93). These cells eat the red corpuscles in the same way that the amoeba eats its food. There are about five million red corpuscles in every cubic millimetre of blood. In addition to the red

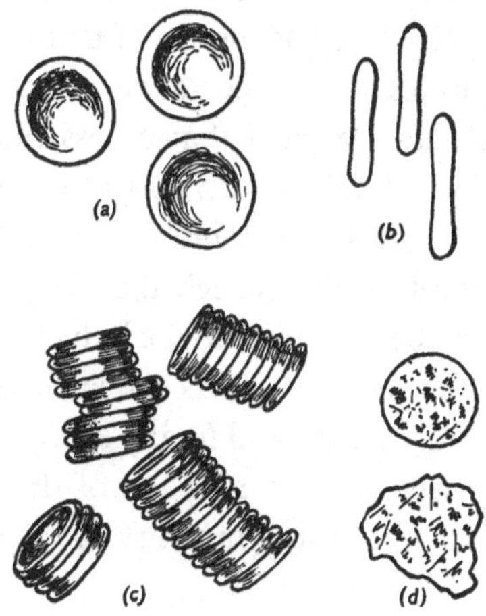

Fig. 98. *Blood corpuscles. (a) Red corpuscles seen flat; (b) on edge; (c) red corpuscles run together in rows; (d) white corpuscles*

corpuscles there are several kinds of WHITE CORPUSCLE. The larger kind is formed in the red bone marrow, and the smaller ones in the lymph nodes (Fig. 98). These cells are very like the amoeba, and are constantly changing shape as they swim about in the plasma. There are also small things called PLATELETS, which are also probably formed in the bone marrow and take part in the clotting of blood.

Work done by the blood

The blood performs many important functions in the body. The red corpuscles carry OXYGEN from the lungs to all parts of the body, and the plasma brings back CARBON DIOXIDE from all parts of the body to the lungs, where it is got rid of.

Food is carried to all parts of the body by the plasma of the blood. In the small intestine digested carbohydrates and proteins pass into the blood plasma, which passes along the portal vein to the liver. Food needed for immediate use is passed into the circulation to be taken to all parts of the body. We have already learned that excess carbohydrate is changed into animal starch or GLYCOGEN and stored in the liver (see page 152). Excess proteins are broken down in the liver to form a simple sugar, which the body can use, and urea which is carried by the blood to the kidneys. On page 159 we read that the digested fat enters the lacteals and not the blood capillaries. These unite and finally enter a blood vessel. Digested fat may be carried to all parts of the body for immediate use or it may be stored in the skin, the muscles or in the body spaces.

As blood flows through the two KIDNEYS (Fig. 99) waste proteins pass out of the blood into very small tubes. In each kidney these tubes unite to form one large tube which is called the URETER. The liquid containing this waste protein matter is called URINE. It passes down the tube, or URETER, from each kidney to the BLADDER, where it is stored. At regular intervals the bladder gets rid of the urine.

Bacteria are killed by the blood. This RESISTANCE TO DISEASE is brought about in several ways. Some white corpuscles eat the bacteria in a similar manner to that in which the amoeba eats its food. The white corpuscles may even squeeze their way through the wall of the blood vessel into the surrounding cells to do so. Some white corpuscles give out substances which kill the germs. The blood plasma also contains substances called ANTIBODIES which are produced chiefly in the liver and lymphatic glands. They neutralize the poisons of the germs or kill the germs. Bacteria are often killed in the

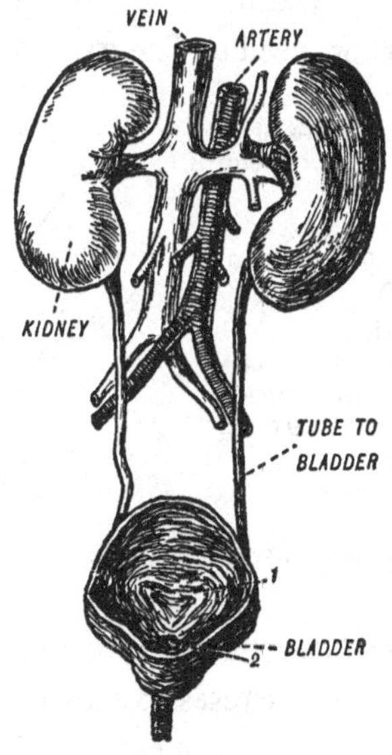

Fig. 99. *The urinary organs.* 1, *Places where the tubes from the kidneys enter the bladder.* 2, *Tube from the bladder*

lymph nodes. The lymph glands in the throat swell when you have a sore throat, as they are getting rid of the germs. If germs enter the body, the neighbouring blood vessels swell up to hold more blood which can fight the germs. This makes the spot hot and inflamed.

Blood carries HORMONES. These are substances which are made in special glands in our bodies, and which greatly affect our health and our temper. Hormones pass directly into the blood stream from the gland. The PANCREAS produces a hormone called INSULIN which controls the amount of sugar that is in the blood. It must therefore affect the liver, where extra sugar is stored. People that have not any insulin in their bodies suffer from a disease known as SUGAR DIABETES. Insulin can be given to them, but as it is destroyed by the juices in the stomach it has to be injected into the blood by means of a hypodermic needle. The THYROID GLAND in the

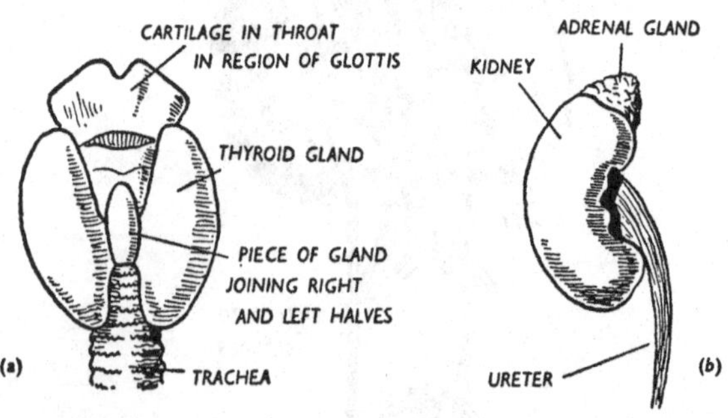

Fig. 100. *Hormone glands.* (*a*) *Thyroid gland;* (*b*) *adrenal gland*

throat (Fig. 100*a*) regulates the rate of mental and of physical activity. Too little thyroid causes loss of hair, thickening of the skin, fatness and very low mental ability. This can be cured by taking thyroid extract through the mouth. Children who lack thyroid are very stunted in growth and are very backward mentally. If too much thyroid is produced the patient is very excitable and nervous and may have bulging eyes. Patients are often hungry even if they eat plenty of food, because their bodies are over-active. No food is stored, so the patient gets thin. Over-production of thyroid can be cured either by killing some of the thyroid cells with X-ray or by cutting away part of the gland. The ADRENAL GLANDS that lie close to the top of each kidney (Fig. 100*b*) are divided into two parts. The thin, red, inner part produces a hormone called ADRENALIN. If too much of this hormone is produced the heart beat increases, the sugar content of the blood goes up, muscular power increases, and the body does not get tired very easily. In fact, it is this gland which, by secreting extra adrenalin, enables you to jump over a gate if a bull is chasing you. The outer, yellowish part or CORTEX produces substances which have been found necessary for life. Work is still being done to find out exactly what part the cortex plays in our lives. The PITUITARY BODY, which is at the base of the brain, affects the growth of the body and also the functioning of the reproductive organs. It also affects or controls all the other glands in the body. The reproductive organs also produce hormones, but these will be dealt with in the next chapter.

BLOOD GROUPS. There are several different kinds of blood in human beings, A, B, AB and O being the main ones. You know that some people called DONORS give up some of their blood in order that it can be given to another person who is very ill. Before the blood from one person is transfused into the body of another it is necessary to know what kind of blood the donor and the patient have. Group O blood can be given to anyone; these are called universal donors. Group A blood can only be given to group A, B to B and AB to AB. If A blood is mixed with blood containing B, or vice versa, the red corpuscles stick together in clusters and cannot do their work, so the patient dies. Group O can only receive O blood, group A can receive A or O, B can receive B or O and group AB can receive AB or O.

The skin

The body is covered with skin which performs several functions. The outer cells are dead, forming a horny layer which is gradually rubbed off. As more living cells become exposed to the air they also die. This outer layer of dead cells prevents the body from losing water, which would otherwise take place since it lives on land. Bacteria cannot

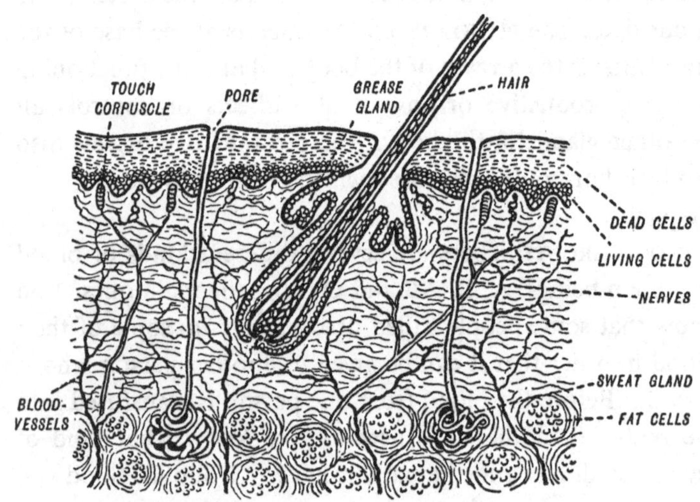

Fig. 101. *Section through human skin*

enter the body through the skin unless it is damaged. If you can look at a section of human skin under a microscope it will look like Fig. 101. The HAIRS grow from cells at the bottom of little pits or hair FOLLICLES in the skin. Small glands called SEBACEOUS glands, which secrete grease, open into these pits. FAT is often stored under the skin. The skin is well supplied with nerves whose endings receive messages of touch, etc. All over the skin there are tiny PORES, which lead to the SWEAT GLANDS which give out the sweat.

TEMPERATURE CONTROL. In cold weather the blood vessels in the skin contract, bringing less warm blood to the surface. In hot weather, or when the body is overheated by exercise, the blood vessels dilate bringing more blood to the skin where it will be cooled. The body also loses heat by evaporating sweat which is produced by the sweat glands and passed to the surface of the skin along the sweat ducts.

Nervous system

The BRAIN, which is protected by the skull, and the SPINAL CORD, which is protected by the backbone, form the CENTRAL NERVOUS SYSTEM. The nervous system is rather like a telephone system throughout the body. The brain and the spinal cord may be compared with the telephone exchange which receives and sends out messages. Small white threads or NERVES send messages from all parts of the body to the central nervous system. These nerves are called SENSORY NERVES. They arouse in our brain the sensations of touch, sight, hearing, pain, etc. Other nerves, called MOTOR NERVES, take messages from the central nervous system to other cells or to muscles in the body. Some MIXED nerves are able to carry messages to and from the brain and spinal cord.

REFLEX ACTIONS. If you prick your finger or touch something that is very hot, a message is sent along the sensory nerve to the spinal cord. A nerve in the spinal cord receives the message at (1) and passes it to (2) (see Fig. 102) where it sends a message along the motor nerve to the muscles in the hand and arm causing them to jerk away. We cannot usually control these reflex actions, as they are called. If you picked up a kettle of boiling water and the handle was very hot the reflex action would not be quite so simple, because if you dropped the kettle you would be scalded. So the brain sends messages to the muscles of your feet and of your arms, enabling you to put the kettle down quickly in a safe place. Think for a moment of all the things that you do every day and you will realize what a wonderful thing your brain is, as it controls every action in your body.

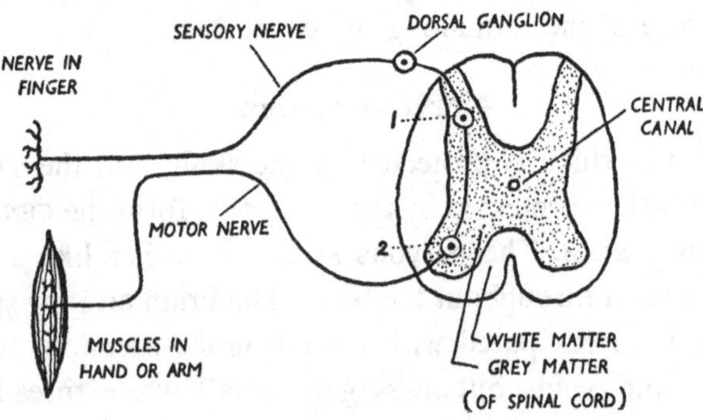

Fig. 102. *Diagram to show how a reflex action works*

Special sense organs

The body contains several different kinds of cell which are able to feel certain sensations. For instance, there are cells in the skin which are sensitive to TOUCH and enable us to feel things.

Touch. The sense of touch is more acute on some parts of the body, such as the lips, tongue, cheeks and finger tips. There are also organs in the skin to feel heat, cold, the amount of pressure, and pain. The sensations of taste, smell, sight and hearing are only felt in certain parts of the body.

Taste. We are able to taste different things by means of the delicate skin or MUCOUS MEMBRANE which covers the TONGUE and the SOFT PALATE, which is the back part of the roof of the mouth. Look at your tongue in a mirror and you will see that it is covered with tiny lumps or PAPILLAE on which are the TASTE BUDS. The taste buds are richly supplied with nerves which enable us to taste our food. The tip of the tongue is sensitive to sweet things, the back to bitter things, and the sides to acid. You can test this for yourself if you touch different parts of your tongue with a glass rod dipped in sweet, bitter or acid liquids such as vinegar. The tongue can also tell whether a substance is hot or cold, hard or soft.

Smell. The MUCOUS MEMBRANE which lines the back of the NOSE is richly supplied with nerve endings that are sensitive to smell. The sensitive areas in your nose are about a quarter of a square inch in size, whereas those of a dog are 10 square inches.

The eyes

Sight. The EYES are the organs of sight. Each eye is held in the eye socket by means of six muscles and the optic nerve. The eyeball is kept in shape and protected by a thick outer coat (called the CHOROID). In front of the eye it is transparent and is called the CORNEA. The cornea is covered with a thin transparent layer (called the CONJUNCTIVA) which is continuous with the skin on the eyelids. The choroid is lined with a pigmented layer (called the SCLEROTIC) which is next to the RETINA or sensitive screen of the eye. The coloured part of the eye, which is called the IRIS, lies beneath the cornea. It has a circular opening in the centre called the PUPIL. The iris expands in bright light making the pupil smaller to prevent too much light from entering the eye. In dull light the iris contracts making the pupil larger to admit as much light as possible (see Fig. 104). Where the choroid and iris meet there is a thickening which contains glands and muscles. The glands produce a fluid which nourishes the eye. The muscles are joined to the lens. By their expansion and contraction the shape of the lens is altered to focus for near or far objects. The lens is short and fat for near objects, and long and thin

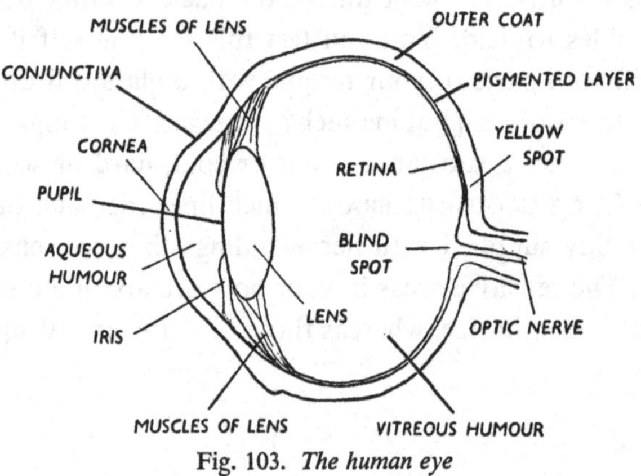

Fig. 103. *The human eye*

for distant objects. Images of objects are produced on a sensitive screen or RETINA at the back of the eye. These images are upside down. The impulses are sent along the

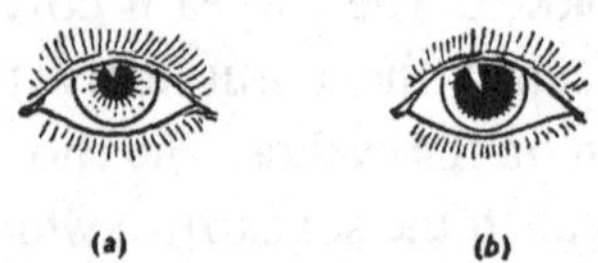

Fig. 104. *The pupil of the eye (a) by day, (b) by night*

optic nerve to the brain, which interprets the images correctly so that we see objects the right way up. Objects on which we concentrate our gaze are focused on the YELLOW SPOT, which is the most sensitive part of the retina. Objects focused on the BLIND SPOT where the nerve enters the eye cannot be seen. This is because the blind spot is insensitive to light. Fig. 105

and the instructions given will show you how to prove the existence of the blind spot.

The eye is not hollow. The space in front of the lens is filled with a watery liquid (called the AQUEOUS HUMOUR) and the chamber behind the lens contains a jelly-like substance (called the VITREOUS HUMOUR).

The eyes are protected from bright light and from dust by the eyelids and eyelashes. There are minute glands under the eyelids which produce a liquid to lubricate them. The tear

X ●

Fig. 105. *The blind spot. Hold the book at arm's length and, closing your left eye, look with the right eye at* X. *Now bring the book slowly closer and you will find that at a distance of about* 8 *in. the black dot will disappear from view. This is because the image falls on the blind spot*

glands produce a watery liquid which keeps the surface of the eye moist and free of dust. Surplus tears pass down a tube into the nose.

The ears

Hearing. Each EAR consists of three parts: the OUTER, the MIDDLE, and the INNER ear.

The OUTER EAR is the part which projects from the head and helps to 'catch' the sound. Have you noticed a cat or a dog moving its ears in order to 'catch' a slight sound? A CANAL which is about 1¼ inches long leads to the middle ear, from which it is separated by the OUTER EAR DRUM (Fig. 106). At the outer end of the canal there are hairs which point outwards, and in the canal there are cells which secrete WAX. The hairs and the wax prevent dust from getting into the ear.

The MIDDLE EAR is almost surrounded by a bony wall. A tube called the EUSTACHIAN TUBE, which is about 1½ inches long, passes from the middle ear to the back of the throat. Air can pass from the throat along this tube to the middle ear, so that the air pressure is the same on both sides of the ear drum. If you have a cold this tube may become blocked, then the air pressure will not be the same on both sides of the ear drum and you will not hear very well. THREE LITTLE BONES, which are joined together to form a kind of a bridge across

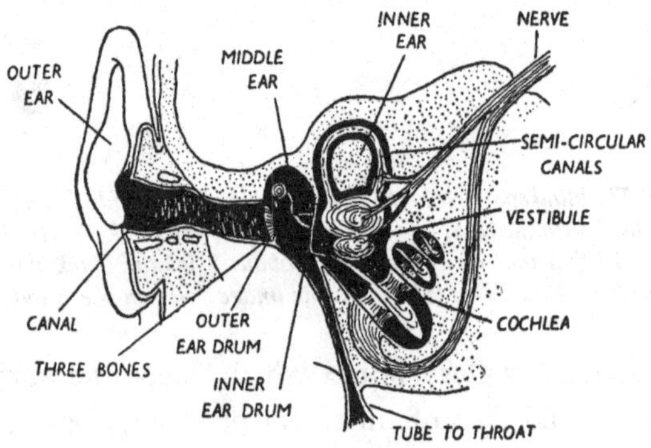

Fig. 106. *Structure of the ear*

the middle ear, form a connexion between the ear drum and the MEMBRANE that separates the middle from the inner ear. This membrane is oval in shape.

The INNER EAR is a complicated cavity that is surrounded by bone. It is divided into two parts. The first part consists of a long spiral cavity called the COCHLEA, which contains the organs of hearing. The impulses which give rise to the sensations of sound are due to the movement of air particles. The SOUND WAVES, as they are called, pass down the tube of the outer ear and cause the ear drum to vibrate. This shakes the three little bones of the middle ear, which in turn shake the oval membrane. The vibration of this membrane moves the liquid which fills the spiral cavity of the cochlea. This movement of the liquid is felt by the nerve endings which take the messages to the brain and so enable us to hear. When the oval membrane moves inwards it exerts a pressure on this liquid, so there is another membrane between the middle and inner ear which is round. This membrane bulges outwards when the oval membrane is pushed inwards, so maintaining constant pressure.

The second part of the inner ear consists of a VESTIBULE and THREE SEMI-CIRCULAR CANALS which are set at right angles to one another. They deal with the position, the balance and the movement of the body. If you move around quickly the liquid in these canals also moves quickly, and when you stop suddenly the liquid continues to move quickly and so you feel giddy. The continual upset of the liquid in these canals causes sea sickness.

CHAPTER 9

CHARACTERISTICS OF LIVING THINGS

Differences between living and non-living things

Now that we have studied many animals and plants it should be easy for us to distinguish living from non-living things. A living thing can feed itself, look after itself and reproduce itself. To do this it can usually MOVE, FEED, GROW, BREATHE, EXCRETE waste matter, REPRODUCE, and is SENSITIVE to STIMULI. Non-living things are not able to perform any of these functions.

Movement

We know that the majority of animals move from place to place in order to get food and to distribute the race. Think of all the animals that you have studied and make a list of the different methods that they have of moving from place to place. The amoeba streams along in the mud; worms wriggle along by the muscular action of the body. Animals that live in water either walk on the weeds, like the caddis larva, or they have some means of swimming. The mayfly larva swims by wriggling its body, whereas other animals have limbs which are specially formed to enable them to swim; for instance, the swimming legs of the shrimp; the strong back and hairy legs of the water boatman; the webbed feet of frogs, ducks and the duck-billed platypus; the fins of the fish; the paddles of a turtle; the wings and webbed feet of a penguin, and the flippers of a whale. Land animals have limbs that enable them to move from place to place. These differ in structure, in number and in size, and are so built that they are suited to the surroundings of the animal. Moles have tiny legs and stout claws to enable them to burrow in the ground; horses have long legs for running in the open country. Some animals have claws for climbing, and other animals (like the monkey) have specially formed hands and feet for grasping. Birds and bats have wings with which they can fly in search of food and to get away from their enemies. Even the young of stationary animals, like the anemone and the barnacle, swim about before they settle down, in order to distribute themselves farther afield.

Some tiny plants have cilia and flagella with which they are able to swim about, but the majority of plants remain in one place. Even these plants are able to move certain parts when they are affected by certain stimuli (see page 186).

Feeding in plants

We have already learned that GREEN PLANTS take in WATER and dissolved mineral salts through their root hairs. The water passes into the leaves where, in the LIGHT, in the presence of CHLOROPHYLL, it joins with the CARBON DIOXIDE that comes in through the pores of the leaf. SUGAR is made, which is the basic food of all living things. If plants are not able to make their own food, they must get food that has already been made. SAPROPHYTES obtain their food from dead things, whilst PARASITES obtain their food from living things.

SAPROPHYTIC plants are usually very useful, because they live on decaying matter and so help to get rid of it. Some bacteria are extremely useful in this way and many fungi, too, live on decaying things. PARASITIC plants live on a living animal or plant, which is called the HOST. We know that there are many parasitic bacteria and fungi living on plants and on animals from which they obtain all their food. These are called TOTAL PARASITES, as they get all their food from their

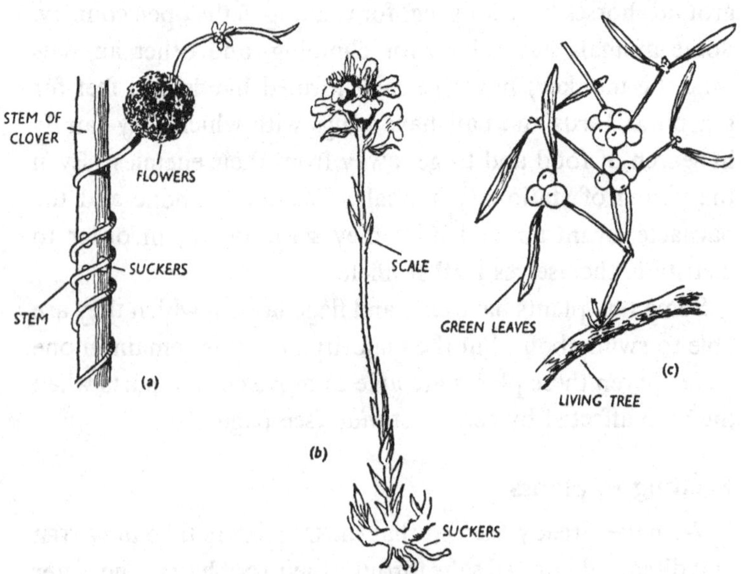

Fig. 107. *Parasitic plants.* (*a*) *Dodder;* (*b*) *broomrape;* (*c*) *mistletoe;* (*a*) *and* (*b*) *are complete parasites*

host. There are several flowering plants that are also total parasites. DODDER is parasitic on clover, lucerne, flax, linseed and gorse. It is usually an annual plant which consists chiefly of reddish or brownish thread-like stems which have only tiny leaves (Fig. 107*a*). The stems twine round the host plant, into the stem of which it sends suckers to obtain food and water. In three months a single plant may kill all the clover over an area of thirty square yards. The plant has tiny, bell-shaped flowers which are white or pink and which grow in clusters. The dodder plant which grows on gorse is a perennial, living for many years. In the autumn only tiny tubercles can be seen where the suckers are. BROOMRAPE also lives on clover. It is an annual plant that consists of a scaly stem that has not any chlorophyll (Fig. 107*b*). It sends suckers into the roots of the clover. It has a spike of reddish, purplish or yellowish brown flowers which produce enormous quantities of tiny seeds.

PARTIAL PARASITES are plants that contain chlorophyll and yet obtain some of their food from other plants. Perhaps the one that you know best is the MISTLETOE, which lives on apple or on oak trees. If a bird gets one of the sticky berries on its beak it may rub it off on the branch of a tree, or if it eats the berry the seeds will pass out with the excreta on to the branch of a tree. Here the seed begins to grow, sending a sucker into the branch to obtain water and mineral salts. YELLOW RATTLE, which is very common in some meadows, is also a partial parasite on the grasses. Although it has green leaves it sends suckers into the host plant.

INSECTIVOROUS PLANTS. Some plants that live in boggy or swampy places are able to make their own food, but they do not get enough nitrates. In order to get more nitrates they catch tiny insects, kill them, digest them and finally absorb the fluids into their own leaves. These plants are called insectivorous plants. SUNDEW is found in peaty, marshy land. It is a small, low-growing plant, with a rosette of round, stalked leaves out of which a number of red tentacles grow (Fig. 108*a*). Each tentacle ends in a club-shaped swelling which is very sticky. If an insect alights on the leaf it is caught by the sticky tentacles. All the tentacles bend over to hold the insect firmly, while glands in the club-shaped tips give out a liquid which dissolves or digests the insect. The juices are absorbed by the leaf and the indigestible parts are blown away when the tentacles release them. BUTTERWORT is also a low-growing plant with a rosette of leaves. These leaves are pointed and are covered with short, sticky hairs. If an insect alights on the leaf it sticks to the hairs. The leaf slowly rolls over from the edges and encloses the insect, which is then

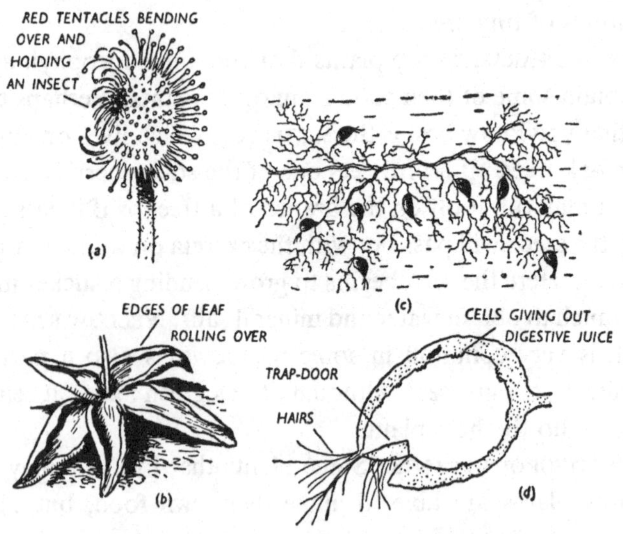

Fig. 108. *Insectivorous plants.* (*a*) *Leaf of sundew;* (*b*) *butterwort;* (*c*) *bladderwort;* (*d*) *one bladder enlarged*

digested. BLADDERWORT lives entirely in water; only the flowers grow above the water. The leaves are very finely divided. Some of them are modified to form BLADDERS (Fig. 108 *c*). Each bladder has a trap-door (Fig. 108 *d*). Tiny water fleas and cyclops can swim into the bladders, but the trap-door prevents them from swimming out again. They are digested inside the bladders.

Feeding in animals

Animals, unlike plants, are not able to make their own food, so they have to eat other animals or plants. Animals can be divided into three groups according to the type of food that they eat: (1) HERBIVORES, that eat vegetable matter. (2) CARNIVORES, that eat other animals. (3) OMNIVORES (like ourselves), which eat either plant or animal food.

Herbivores. These animals have to search for their food, but they do not have to catch it. They have jaws and teeth which are specially formed to enable them to eat their food. All members of the snail family have rough tongues with which they file away leaves. Caterpillars and other plant-eating insects have hard jaws to bite leaves or twigs or even the bark of a tree. Birds have special beaks for picking up seed or for cracking nuts. Hoofed mammals and gnawing mammals have square, ridged grinding teeth for chewing their food. Both male and female reindeer have antlers which may be used in searching for reindeer moss in the snow.

Carnivores. Their bodies are so formed that they can catch and kill their prey. They may have limbs which enable them to swim, run or fly after their prey. Animals like the spider make webs in which they trap their prey. Other animals lie in wait for their prey. These animals are often coloured to match the surroundings so that they will not be seen. Many carnivores have sharp, pointed teeth with which they tear and chew the flesh of their prey. Ladybirds and their larvae have hard jaws for devouring smaller animals. The ant-eater has a very long, pointed snout and a long, needle-like tongue for licking ants out of the cracks of trees. It can make these cracks wider with its long, strong claws. Woodpeckers also have needle-like tongues for the same purpose. Birds, fishes and snakes often swallow their prey whole, and all but the birds have small, sharp teeth, which point backwards and are used to hold the prey. Birds have beaks which are specially shaped to enable them to pick up the food that they like (see Book I). Animals may have some means of holding their prey; for instance, the pincer-like front legs of the water scorpion and the claws of many vertebrates.

Parasites. Many animals are parasites on animals or plants. EXTERNAL PARASITES are those which remain on the outside

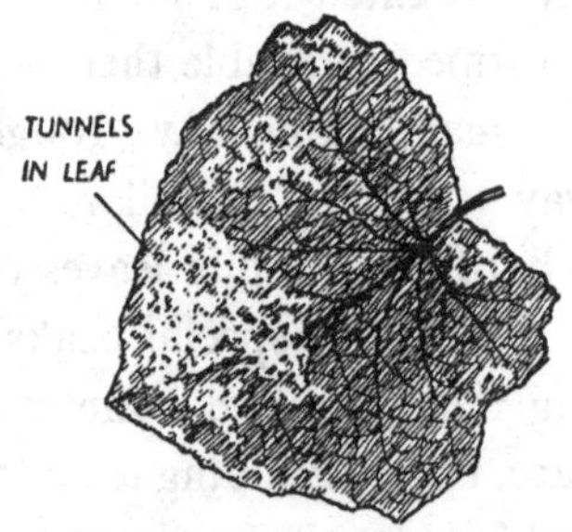

Fig. 109. *Leaf miner*

of the host's body. Some external parasites, such as the lice, mites, ticks and wingless aphides, always remain on the host's body. Gnats, mosquitoes, bugs, fleas and horse leeches only come to their host when they wish to feed. All these animals have special mouth-parts, which enable them to pierce the skin of the host and to suck the blood or the sap. They also have some means of clinging to the host; the horse leech has a sucker round its mouth and lice have claws. INTERNAL PARASITES live inside the body of the host. The leaf miner lives between the upper and lower surfaces of a leaf (Fig. 109). The ichneumon-fly larva lives inside a caterpillar, which it gradually kills. Animals such as the tapeworm live in the food canal, where they are surrounded by digested food which they absorb through the surface of their body. Try to make a much longer list than this of all the various ways by which animals procure their food.

Growth

All living things grow. This means that they take in food which they change into living material in their body. Plants continue to grow throughout their lives and have no definite size. Animals only grow to a limited size.

Respiration

All living things respire. They take in OXYGEN, which is sent to all parts of the body. The oxygen is needed to break down the foodstuffs which have been made and stored in the animal's or plant's body. When the foodstuff is broken down ENERGY, which was stored in the food, is released. Energy is necessary for all the living processes in a living thing. As the food is broken down CARBON DIOXIDE is formed which must be got rid of.

In very simple living things the gases pass through or diffuse through the surface. When animals or plants are more complicated in structure they have special methods of taking in the oxygen and passing it to all parts of the body, and of collecting and getting rid of the carbon dioxide. The shoots of PLANTS breathe through the pores which are on the leaves and young stems, or through the lenticels in older stems. The oxygen and carbon dioxide pass along the spaces between the cells. The roots of plants breathe through the root hairs. Oxygen, which is dissoved in the water in the soil, passes into the root hairs, and carbon dioxide diffuses out. ANIMALS have many ways of taking in oxygen and giving out carbon dioxide. The diffusion of gases takes place through the skin of a worm;

snails and slugs have a breathing hole which leads into a lung that is like a skin bag. Insects, millipedes and centipedes have spiracles which lead into air tubes, and spiders have book lungs. Many water animals have gills which vary very much in structure; for example, the gills of the swan mussel, the gills which grow out of the side of the body in the mayfly and caddis-fly larvae, the external and internal gills of tadpoles, and the internal gills of fish. Amphibians, when adult, reptiles, birds and mammals have breathing holes which lead to the lungs. Oxygen is taken to all parts of the body in air tubes or in the blood.

Excretion

All living things are able to get rid of any waste matter that they do not want. Make a list of the ways in which waste matter is excreted in the plants and animals that you have studied.

Sensitivity

All living things are affected by their surroundings. Animals have special sense cells which make them aware of their surroundings. These may be only single cells or small groups of cells in the skin which are sensitive to touch, warmth, light and chemicals, or they may form special sense organs. Some examples of sense cells and organs are the skin of a worm, the sensitive skin on a bat's face, feelers, whiskers, tongues, eyes, nose and ears.

Plants are also sensitive to certain stimuli. Some flowers, such as the dandelion, open in the sunshine and shut in the rain. Cones open in fine weather to scatter the seeds, and close in wet weather when the winged seeds cannot be blown about. The leaves of sorrel and clover and flowers of dandelion close at night, and the leaves of mimosa move when they are touched (Fig. 110). The shoots of plants always grow towards the light. You can show that shoots grow towards the light if you place germinating seedlings in a box which has a hole at one end (see Fig. 112).

The force due to gravity affects the growth of plants. Roots grow towards the stimulus and shoots away from it.

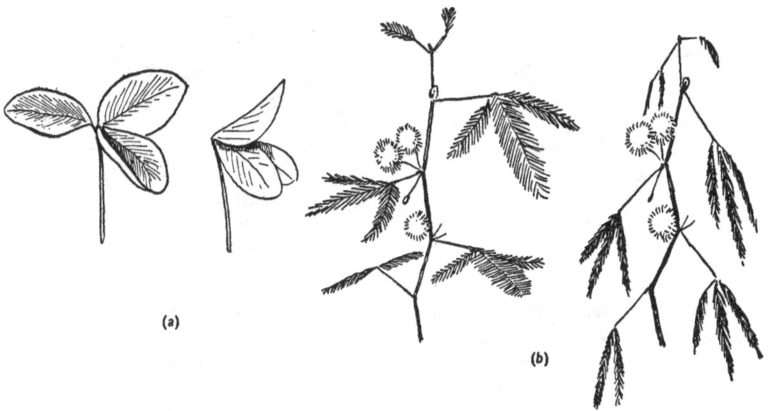

Fig. 110. *Plant movements.* (*a*) *Day and night positions of clover leaf;* (*b*) *mimosa leaves before and after they have been touched*

You can demonstrate this in the following way. Place seeds in a jar containing damp blotting-paper (see Book I, Fig. 111). Leave them until the root and shoot have grown a little. Remove one seedling and replace it after turning it upside-down. You will see that the root bends as it grows until it is pointing downwards again. Similarly the shoot bends as it grows and points upwards again. The leaves of insectivorous plants move to hold insects. Leaves will always move to face the light. This you can see if you put a potted plant in the window. The leaves will move to face the light. Move the

Fig. 111. *More plant movements. Privet branches. (a) Normal arrangement of leaves in pairs alternately down the stem; (b) similar twig from side of hedge. The leaf stalks have twisted so that the leaves can get the greatest amount of light*

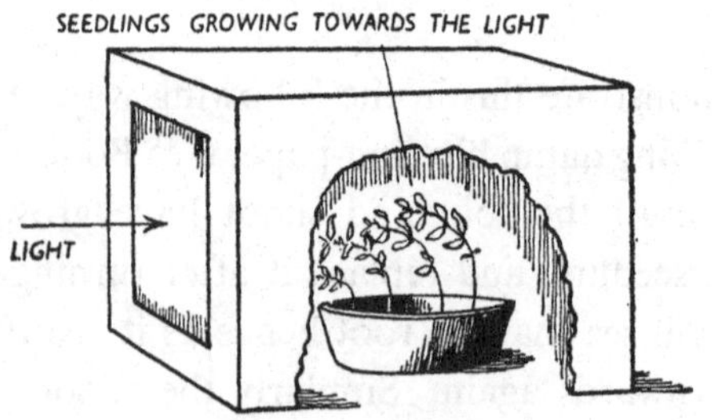

Fig. 112. *Experiment to show effect of light on direction of growth in plants*

plant round again and you will see the leaves turning once more. Some climbing plants have stems which twine around an object when they feel it.

Reproduction

Nearly all living things die at a certain age, and so they must leave young ones to take their place. This production of young animals and plants is called REPRODUCTION. Reproduction takes place in several ways.

(1) Many single-celled animals and plants, such as the amoeba, simply split into two when they reach a certain size. In these living things there are not any parents to grow old or to die. This is called ASEXUAL REPRODUCTION.

(2) Pieces of animals and plants may develop into new individuals. Hydra and yeast develop buds which grow into new individuals, which become separate from the parent. Cuttings of many plants, such as carnations, blackcurrant bushes, willow and poplar, will grow into new plants. Parts of plants also grow into new plants; for example, the bulbs of daffodils and onions, crocus corms, irish rhizomes and potato tubers. This is called VEGETATIVE REPRODUCTION.

(3) It is, however, usual for new individuals to grow from some kind of EGG CELL which is produced by the mother or FEMALE. The eggs of the stick insect and the summer eggs of water fleas and aphides grow into new individuals right away. In the majority of living things the eggs cannot develop until they have been fertilized. In Book I we learned how the egg cells of plants, which are called OVULES, are fertilized by the pollen grains, and the eggs of hydra are fertilized by the sperms. All egg cells in plants and in animals are fertilized in this way; that is, the nucleus of the egg cell joins with the nucleus of the pollen grain, or as it is called in animals,

the SPERM CELL. Pollen grains and sperm cells are the father or MALE cells. This is called SEXUAL REPRODUCTION.

In many animals and plants the female and the male cells are produced in the same individual. These are said to be HERMAPHRODITE. In spite of this, very few of them fertilize themselves, because the offspring are stronger if the egg cells are fertilized with the pollen grains or the sperm cells of another individual. Many hermaphrodite animals and plants are not able to fertilize themselves. For instance, in some flowers the stamens and the pistils ripen at different times. The stamens of the willow-herb ripen before the ovules, and the ovules of the coltsfoot, horse chestnut and arum lily ripen before the pollen grains. These plants must be cross-fertilized with the pollen of another plant, which is carried to the plant by wind or by insects. The pin-eyed primrose cannot fertilize itself because the stamens are below the pistil. Although worms and snails are hermaphrodite, they also fertilize one another.

In order to make sure that they are cross-fertilized many living things are UNISEXUAL. This means that the ovules and the pollen grains or sperm cells are produced by different individuals. The stamens and the pistils of dog's mercury, willow, poplar, stinging nettle, hop and many more plants grow in different individuals, and they rely on insects or the wind to pollinate them. Many animals are unisexual, and as the eggs and sperm cells are formed inside the animal it is obvious that they cannot be fertilized in the same way that plants are. In some animals that live in water, such as the hydra, the sperms are shed into the water. They swim by means of a little tail to the egg cells, to which they are attracted by a chemical substance which is given out by each egg cell. In fishes such as the herrings, which live in shoals, millions of eggs and sperms are passed into the water, where the eggs are fertilized. You have probably eaten herring roes. The hard roes are the eggs and the soft roes are the sperms. The male stickleback makes a nest of weeds, inside which one or two female sticklebacks lay their eggs. The male stickleback

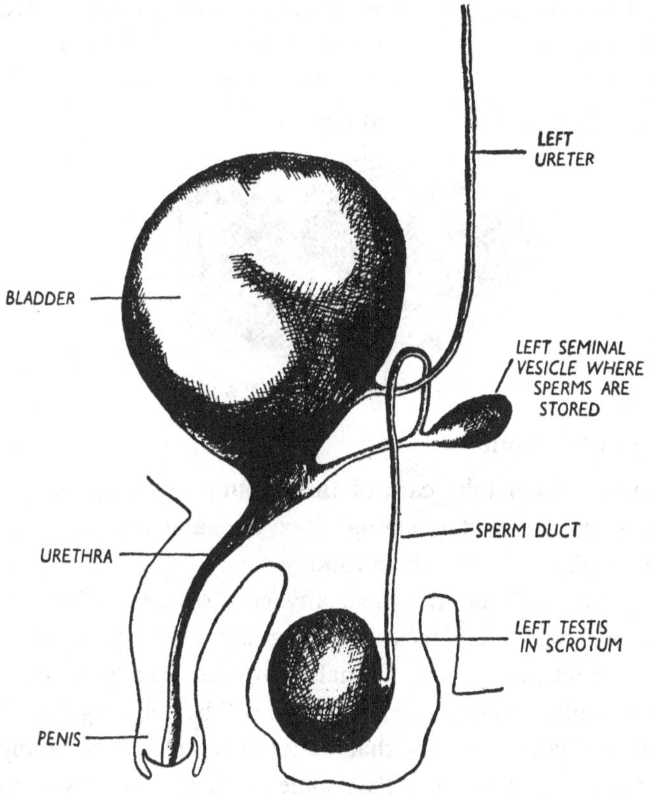

Fig. 113. *Male reproductive organs, left side*

then sheds the sperm cells on top of them. A male frog or toad sits on a female's back to shed its sperms on to the eggs as the female lays them in the water. Dogfish eggs are protected by a case which is made by a substance that is secreted in the female's body (Fig. 51). Birds' and reptiles' eggs are also covered with a shell when they are laid. These eggs obviously cannot be fertilized after they are laid. The eggs are fertilized inside the female's body before they are surrounded by the shell. The eggs of all land animals, such as reptiles, birds and mammals, like those of dogfish, have to be fertilized before they are laid. The sperm cells are placed inside the female's body by means of a special organ called the PENIS. The sperm cells then swim to the egg cells and fertilize them.

Fig. 114. *Midwife toad carrying eggs*

Care of the young

Plants do not take care of their young ones. The seeds, when scattered, contain enough food to enable them to grow big enough to get their own food.

Some animals do not take any care of their offspring. Many insects, worms, fishes and frogs lay their eggs and then forget about them. These animals, however, usually lay their eggs in a place where the young ones will be able to get food when they hatch. Animals that do not take care of their young lay many eggs to ensure that some of them will grow into adult animals.

Woodlice, prawns and some toads carry their eggs around with them until they hatch (Fig. 114). The male stickleback guards the eggs in the nest until they hatch, and then he protects the young for a time. Birds and social insects lay their eggs in nests or in hives. The eggs are protected until they hatch, and the young ones are fed until they are big enough to feed themselves. The eggs of the adder, some fish such as the blennies, and the summer eggs of water fleas and aphides, hatch before they are laid. As soon as the young are born they have to fend for themselves.

The eggs of all the animals that we have mentioned so far contain enough food to feed the baby animal until it hatches out of the egg. Mammals have tiny little eggs which contain

Fig. 115. *A blennie*

very little food. Sperms which are put into the vagina (see Fig. 116) of the female by the male, swim through the uterus to the oviducts where the eggs are fertilized. When the eggs are fertilized they bury themselves in the wall of the UTERUS of the mother (Fig. 116). The wall of the uterus thickens and a structure known as the PLACENTA develops between the developing egg and the wall of the uterus. Oxygen and food pass from the blood of the mother into the cells of the placenta. From here the oxygen and the food pass into the blood of the developing animal or EMBRYO. Each embryo floats in fluid which is enclosed in a skin-like bag. This protects the embryo if the mother receives a blow. Similarly, waste products pass from the blood of the young animal into the placenta and then into the mother's blood. As the young animal gets bigger it is connected to the placenta by a cord, which carries two arteries and one vein. Baby kangaroos are only about one inch long when they are born, and young rabbits, mice, hedgehogs, monkeys and even human babies are not able to look after themselves when they are born. Young horses, cows and other hoofed mammals can walk when they are born, but they cannot feed themselves. If the young are immature when they are born they are either kept in a pouch on the mother's body or in a nest which is made by the mother.

All young mammals and many young birds are unable to feed themselves at first. Mother and father birds may partly

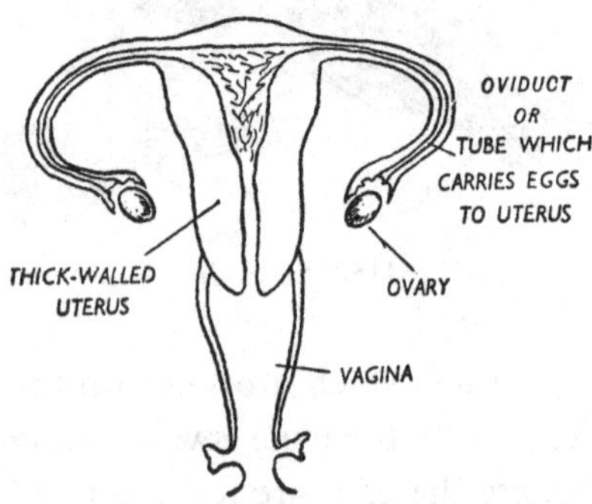

Fig. 116. *Female reproductive organs*

digest their food and then bring it up into their mouths and put it into the mouths of the baby birds; or they may find food and place it into their babies' mouths. If you look at a baby bird that is fed in this way you will notice that it has a very wide mouth. Baby mammals, however, are fed on milk which is produced in special glands in the mother's body. Mammal parents take care of their young ones until they are able to fend for themselves. The parents of many young animals teach their offspring how to do all the things that they must know how to do, such as fly, swim, protect themselves, and find or catch their food.

CHAPTER 10

EVOLUTION

Formation of the earth

The crust of the earth is divided into layers which are called STRATA. The oldest strata are, of course, at the bottom. It is thought that the earth was once a molten mass which gradually cooled. Those rocks which were formed by heat were named IGNEOUS ROCKS. The later rocks were named SEDIMENTARY ROCKS because they were formed in the following way. As the rivers flowed down to the sea they took with them large quantities of mud, which was either deposited round the mouth of the river, or on flooded plains, or it was swept out to sea. The particles of mud gradually sank to the bottom. As more and more sediment was laid down, the mud or sand beneath was turned into hard rock by the pressure of the layers above. Similar layers may have been formed in pools and lakes, and also in peat country where plants, as they died, formed layers on top of plants that had been dead for some time. Volcanoes, when they are active, send out rock fragments, dust and lava, which also form a layer on the earth that surrounds the volcano. As all these layers were being formed, the bodies of dead animals and plants were buried in them. The softer parts of these dead bodies were probably destroyed by bacteria, but the harder parts remained in the ground and were changed into FOSSILS.

7-2

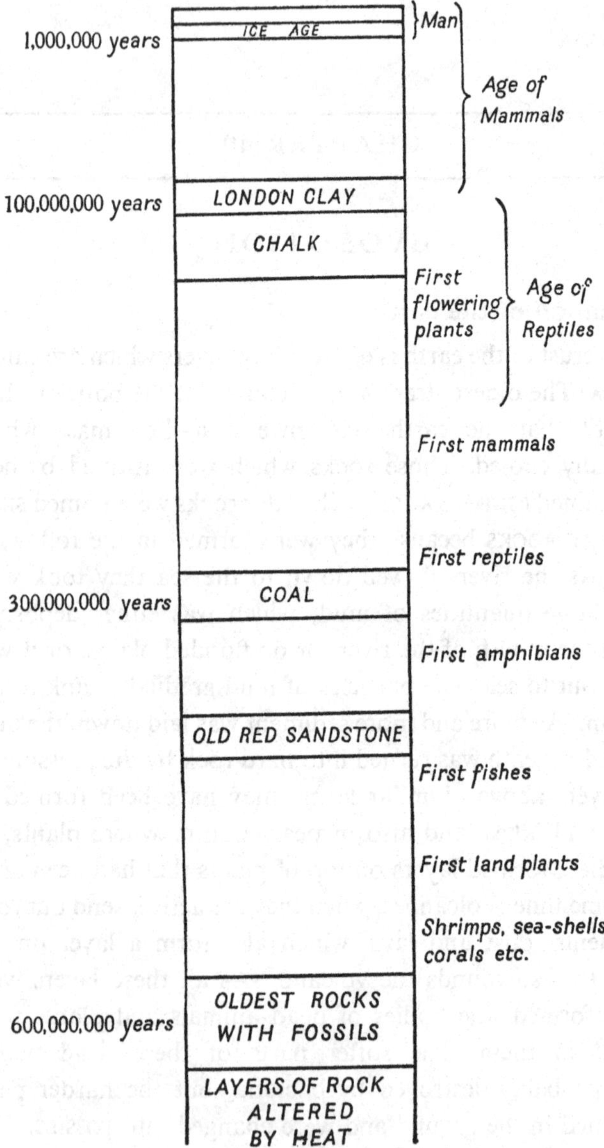

Fig. 117. *Table showing the layers of rocks formed under the sea, with their ages and the chief fossils found in them*

Movements of the earth

We know that the centre of the earth is a molten mass. We have evidence of this when a volcano is active. It is easy then to understand that the crust of the earth very gradually moves up and down on the top of this molten mass. Land which was once below the sea may now be above it, and vice versa. In some parts of England you will find the fossils of sea animals; this is so even in the middle of England, which shows that the land was once below the sea. In spite of these movements in all parts of the world it has been found that the different layers, each of which probably took millions of years to form, are in the same order (see Fig. 117). Each of these layers has its own particular kinds of fossil. These show that certain types of animal and plant were living at different periods in the world's history.

The world in past ages

When the earth was first formed there was no life on it. The first living things to appear were very simple creatures, probably as simple as the amoeba. These were followed by animals similar to worms, shrimps, molluscs and corals. No fossils have been found of the soft-bodied animals, since they rotted away as they died. In more recent layers than those in which these animals appeared, the remains of fishes and of land plants have been found. Many of the first land plants had leaves similar to the ferns, but they grew into large fern trees. It was these trees, which were living 300,000,000 years ago, that have been gradually changed into coal. Later, amphibians and reptiles appeared which were not at all like the reptiles of today. Many of them were huge animals and some of them could fly. Some amphibians that were similar to the present-day newt were 9 feet long. The fossil skeletons of some reptiles may be seen in the Natural History Museum in London. At this time coniferous trees began to appear. Later came mammals, which were very small and different from those of today. In the more recently formed layers mammals, birds, flowering plants and trees gradually appeared. If you arranged all living things in order, beginning with the simplest and gradually going to the more complicated, you would find that this arrangement would correspond to the order in which living things appeared on the earth.

Origin of life on the earth

Many theories have been put forward suggesting how life began on the earth, but we have no proof of what the first living things looked like. All we know is that today all living things come from living things, but originally life must have begun from something that was not alive. It is thought, too, that life began in the water, and that the first living things were simple in structure. Many of the oldest fossils are of water animals. Nearly all the simplest animals and plants that are alive today live in water.

Gradual change from life in water to life on land

If you think for a moment of all the groups of animals and plants, beginning with the simplest, you will realize that the structure of all living things gradually became more complicated to fit them to live on the land. The more primitive plants, such as algae, fungi and mosses, cannot live or survive without water because they soon dry up. The sperms of mosses and ferns have to swim in water to reach the egg cells before new plants can be formed. As plants adapted themselves to life on the land they had to prevent themselves from drying up, and have alternative means of reproducing themselves. In addition, as plants became taller they had to develop some means of supporting themselves.

Many primitive animals also live in the water, including some of the vertebrates. Fishes live entirely in water, and amphibians have to return to the water to breed. Many animals have learned to live on the land and even in the air. Their structure has become more complicated to prevent them from drying up, to enable them to breathe on land, and to move about in search of food. The method of reproduction has changed to suit the environment in which they live.

Plant evolution

The simplest plants, such as the algae and the fungi, live in water or very damp places, and must be very similar to their ancestors. Mosses have tried to live on the land, but they must have a certain amount of water to prevent them from drying up and to enable the sperms to swim to the egg cells. Ferns can live in drier places and only need water for their sexual reproduction (see page 41). Flowering plants have conquered the land in two ways. First, they have learned to prevent loss of water in various ways: (1) They lose their leaves in the winter. (2) They form thick cuticles. (3) All aerial parts die down in the winter. (4) Leaves are reduced in size, as they are in gorse. (5) Leaves are absent, their work being carried on by the stems. (6) The leaves may be rolled. Secondly, they have pollen grains instead of sperms, which are carried to the pistil by insects or by wind. The seeds, too, are scattered by wind, animals or explosive mechanisms.

Fig. 118. *A fossil leaf of a fern which lived in a coal forest. It was alive 300 million years ago*

Animal evolution

The oldest fossils that have been found are those of corals, whelks, mussels, shrimps, etc. They are similar to, but not exactly like, those that are living today. The earth was probably populated before this time with animals which were more simple in structure; but no fossils of them have been found, for two reasons: (1) their bodies were probably too soft to

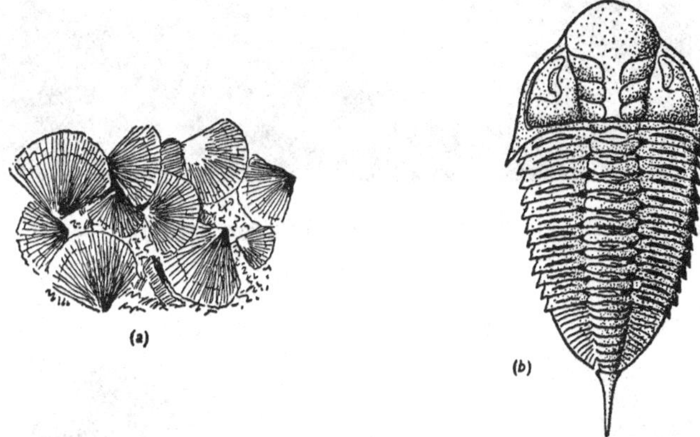

Fig. 119. *Fossil invertebrates.* (*a*) *Brachiopod shells;* (*b*) *a trilobite*

change into fossils; (2) we know that the oldest rocks have been pushed down towards the centre of the earth and have been heated and completely changed. So even if there had been any fossils, they would have been destroyed. Fishes were the next animals to appear and then insects, amphibians, reptiles, and mammals and birds that were unlike those of today. Finally, modern mammals and birds and man appeared. Some of the older birds had teeth similar to those of the reptiles. The reptiles were large clumsy animals. We can tell by the teeth and the skeletons that have been found

Fig. 120. *Two kinds of flying reptile as they must have appeared*

Fig. 121. *What a giant extinct reptile looked like that lived 100 million years ago. It was eighteen feet high and was a beast of prey*

that the ancient reptiles were either animal or vegetable feeders, even as the mammals are today. They had very tiny brains, which probably explains why they did not survive. The first mammals were very small.

Do not think that reptiles, mammals or birds suddenly appeared on the earth from nowhere. Some animals gradually changed in structure over millions of years, to adapt themselves to the change of surroundings or of climate. Many fossil skeletons of horses have been found which

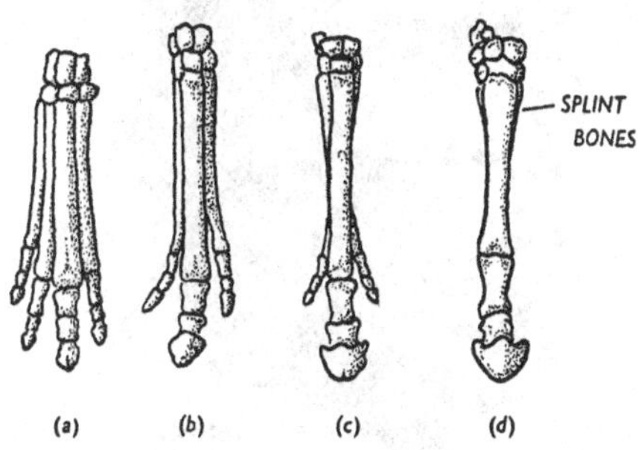

Fig. 122. *Evolution in feet of horse.* (*a*) *Foot of a fossil horse which lived at the time that London clay was formed;* (*b*) *and* (*c*) *later fossil horses;* (*d*) *a horse of today*

illustrate how these changes have come about. Horses were originally small animals with four toes on each foot. The skeletons of horses that lived millions of years later had only three toes. As the horses became bigger they walked only on the middle toe, and the two outer toes gradually became smaller and smaller until they appear today as tiny splint bones (Fig. 122).

Fossil men

Human beings have been on the earth for about a million years. In 1892 the remains of an ape-man were found in Java.

8-2

This man formed a link between the man-like apes and human beings. He had a small brain, huge brow ridges over his eyes like those of a gorilla, but he walked upright. Fossils of several kinds of men have been found in different parts of the world, and they all show that ancient men had small brains, massive jaws and large canine teeth.

Fig. 123. *The Java ape-man*

Causes of evolution

We have many evidences from the fossils and from the structure of plants and animals that evolution has taken place, but you may ask how evolution has occurred.

Heredity

Much work has been done to show that an offspring acquires its characteristics from its parents. When breeding animals and plants we can usually tell what the offspring will be like if we know all the characteristics of the parents, grandparents, etc. Children could not all be exactly alike in one family unless the parents were alike (which is very unlikely in human beings, but which is so in many animals and plants). So we get variations in height, hair colour, colour of the eyes, shape of the nose, etc. These VARIATIONS are due to the different characters that have been acquired from the parents; they are, in fact, due to HEREDITY. Sometimes the environment will cause variations. If one child is well fed it will grow bigger than one that is badly nourished. Sometimes, however, new variations appear which cannot be accounted for by heredity or by environment. These variations, or MUTATIONS as they are called, are passed on to their offspring.

The struggle for existence

All living things are continually fighting to live. The weaker ones die and those that are the stronger live. This is called NATURAL SELECTION. These animals and plants will acquire characteristics which fit them to their mode of life. It is only those living things which change with the ever-changing surroundings that survive. The variations so acquired will be handed on to their offspring. So in time the descendants will gradually change. Today we cannot see this happening in a country like ours, as everything is done to help the weak people to live.

QUESTIONS

When possible answers should be illustrated with clearly labelled diagrams

Chapter 1

1. What are the main differences between animals and plants?

2. Is it true to say that animals could not live without plants? Give reasons for your answer.

3. What are the two main divisions in (*a*) the plant world, (*b*) the animal kingdom?

Chapter 2

1. Briefly describe, with the help of diagrams, how (*a*) spirogyra, and (*b*) brown seaweed reproduces.

2. Where could you find brown seaweed? Explain why it can grow in this place.

3. Fungi have not any chlorophyll so they cannot make their own food. How do they obtain their food?

4. Give a brief account of some of the damage done by parasitic fungi.

5. With the help of diagrams, describe how pin mould reproduces, (*a*) when it has plenty of food, (*b*) when the food is nearly used up.

6. What uses do we make of yeast?

7. What are bacteria? To what extent are they (*a*) useful, (*b*) harmful, to us?

8. How are virus diseases spread from one plant to another?

9. What harm can viruses do to plants?

10. Name four diseases in animals that are caused by viruses.

11. What is a lichen?

12. Moss plants do not have real seeds. Explain as briefly as possible how new moss plants are formed.

13. What are the brown patches on the underside of fern leaves?

Chapter 3

1. How could you recognize the following trees: (*a*) pine, (*b*) fir, (*c*) larch, (*d*) cedar?

2. Of what use are the cones of a coniferous tree?

3. What are the two main groups of flowering plants and how would you recognize members of each group?

Chapter 4

1. Compare and contrast the anemone and the hydra.

2. How does a jellyfish reproduce?

3. How does (*a*) a starfish, (*b*) a sea urchin move from place to place?

4. Briefly describe a flat-worm. Where do they live?

5. Describe the life history of a liver fluke.

6. How could you prevent the sheep from getting liver flukes?

7. Describe the life history of a tapeworm.

8. What are the three main groups of worms? How would you recognize members of each group?

9. How does (*a*) a pond snail, (*b*) a sea snail breathe?

10. How is it that the molluscs that live in the sea are not washed away?

11. Many bivalve molluscs bury themselves in the sand. How do they do this? How do they breathe when they are buried?

12. How is a snail's shell formed?

13. How does a crab differ from a lobster?

14. What are the chief characteristics of the arthropods?

15. Describe any animal that belongs to the Crustacea family. Point out the characteristics of the family.

16. How does a barnacle protect itself? How does it get its food?

17. Compare and contrast a spider and an insect.

18. Write a brief account of the harm that is done by mites and by ticks.

19. Where does a water spider lay its eggs?

20. How would you recognize an insect?

21. Give *one* feature only that will enable you to recognize the following insects: (*a*) bug, (*b*) mayfly, (*c*) beetle, (*d*) house-fly.

22. How does the life history of a froghopper differ from that of a fly?

23. Write a brief essay on the harmful insects that live in a house.

24. Name two beetles that are harmful, and explain what damage they do.

25. Describe two insects that are useful in the garden.

26. Describe how the mosquito and the fly feed. Explain why their methods of feeding are harmful to us.

27. Write all that you can about fleas.

28. How could you recognize the larva of a sawfly? How does it differ from a caterpillar?

Chapter 5

1. Name the chief groups of vertebrates. How can you recognize the animals that belong to these groups?

2. Explain why a fish has not any eyelids and a bird has three eyelids.

3. What is a flat fish?

4. Describe the life history of either an eel or a salmon. What is unusual about these fish?

5. Explain why it is that a herring lays millions of eggs and a stickleback lays only a few eggs.

6. What are the chief differences between snakes and lizards?

7. Write a few lines about the chameleon or the crocodile.

8. How do the following animals protect themselves: (*a*) tortoise, (*b*) snake, (*c*) crocodile?

9. Penguins swim and ostriches run. What differences has this made to the structure of these birds?

10. How can we tell that a duck-billed platypus is a mammal?

11. How do the following animals move from place to place: (*a*) snake, (*b*) kangaroo, (*c*) whale, (*d*) monkey, (*e*) horse?

12. What care do the following animals take of their young when they are born: (*a*) kangaroo, (*b*) horse, (*c*) mouse, (*d*) bat, (*e*) monkey?

13. Give reasons for the following: (*a*) solid hoof of a horse; (*b*) split hoof of a pig; (*c*) hump of a camel; (*d*) pockets in the stomach of the camel; (*e*) long neck of a giraffe.

14. Write all that you can about the elephant.

15. How are whales adapted to life in the water?

16. By looking at its teeth how can you tell whether an animal is: (*a*) a beast of prey; (*b*) an insect-eater; (*c*) a gnawing mammal?

17. Compare and contrast a monkey or an ape with yourself.

18. How does the wing of a bat differ from that of a bird?

19. What are the differences between the toothed whales and the whalebone whales?

Chapter 6

1. What is the difference between soil and subsoil?

2. Briefly describe how soil is formed.

3. Give the differences between clay and sand. Illustrate your answer by describing any experiment that you have done.

4. What is meant when we say that soil is waterlogged?

5. How can you puddle clay?

6. Briefly describe how peat is formed.

7. What is the best type of soil for plant growth?

8. Give the chief factors that affect the soil.

9. How could you show that soil which appears dry does contain water?

10. For what purpose are water culture experiments carried out?

11. Explain carefully how you would set up a water culture experiment and give reasons for everything that you do.

12. When is soil (*a*) rolled, (*b*) raked or harrowed? Give reasons.

13. Why do we manure soil?

14. What is the difference between organic manures and fertilizers?

15. How is a compost heap made?

16. What is meant by the 'rotation of crops'?

17. Why is lime added to soil?

Chapter 7

1. What is meant by 'transpiration'?

2. Why do the roots of plants take in such an enormous amount of water?

3. Describe an experiment that would show that leaves give out water.

4. Write a brief essay on 'How plants prevent loss of water by the leaves'. Illustrate your answer with the help of diagrams.

5. What is meant by (*a*) deciduous, (*b*) evergreen, plants?

6. How does the wind alter the shape of a tree?

7. What is meant by the word 'photosynthesis'?

8. What conditions are necessary before a leaf can make sugar?

9. Briefly describe how you would show that a leaf contains starch.

10. Describe experiments that you could do to show that the following things are necessary for the formation of starch in a leaf: (*a*) chlorophyll, (*b*) carbon dioxide, (*c*) light, (*d*) water.

11. How could you show that leaves give out oxygen in the day-time?

12. Describe the tests that you could do to find out whether the food stored in a plant was (*a*) starch, (*b*) glucose, (*c*) cane sugar, (*d*) protein, (*e*) fat.

13. Briefly describe how plants respire.

14. Describe an experiment that you could do to show that plants respire.

Chapter 8

1. What are the chief uses of a skeleton?

2. Explain briefly how we are able to bend our backs.

3. Compare the structure of a leg with that of an arm.

4. What is meant by voluntary and involuntary muscles?

5. Describe how the biceps muscle of your arm works.

6. How many teeth do we have in our milk, and in our permanent, sets? Name them.

7. Write a brief essay on foods and their different values in our diet.

8. How would you show that food contained: (*a*) starch, (*b*) sugar, (*c*) protein?

9. What are vitamins? Name the best-known vitamins and say why they are so important in our diet.

10. With the help of a diagram describe the digestive system. Explain how food is digested.

11. Explain what happens to the following foods after they have been digested: (*a*) carbohydrates, (*b*) proteins, (*c*) fats.

12. Describe (*a*) how air gets into the lungs, (*b*) how air is forced out of the lungs.

13. Why should you breathe through your nose?

14. Describe the course of the blood from the left auricle until it returns there.

15. What makes the blood move along the arteries and the veins?

16. Write a few lines about the lymph system.

17. Describe the structure of blood.

18. What are the functions of the blood?

19. How and where does oxygen enter the blood?

20. Why does the body need oxygen?

21. How does the blood get rid of waste material?

22. Write all that you know about hormones.

23. Describe the structure of the human skin.

24. What functions are carried out by the skin?

25. How is the temperature of the body controlled?

26. Of what use is the nervous system?

27. How does a reflex action work?

28. How do we taste and smell our food?

29. How are our eyes protected from (*a*) bright light, (*b*) dust?

30. Explain how we are able to see objects (*a*) near to us, (*b*) at a distance.

31. What is meant by (*a*) the yellow spot, (*b*) the blind spot, of the eye?

32. How is sound carried to the nerves of the ears?

33. How is the same air pressure maintained on either side of the ear drum?

34. What other purpose, besides hearing, is served by the ears?

Chapter 9

1. Is it true to say that animals move from place to place and plants do not? Give reasons for your answer.

2. What are the chief differences between living and non-living things?

3. Write an essay on 'Movement in animals'. In your answer you should show how the animals are adapted to their surroundings.

4. What is a parasite? Say something about two of the following: (*a*) dodder, (*b*) broomrape, (*c*) mistletoe.

5. Describe two plants that obtain their nitrates by catching insects. Explain how the insects are caught.

6. What are the differences between a carnivorous and a herbivorous animal?

7. Write an essay on the various ways by which animals get their food.

8. Explain how a green plant makes its food.

9. Describe two parasitic animals.

10. How do the following animals breathe: (*a*) spider, (*b*) caterpillar, (*c*) fish, (*d*) mouse, (*e*) earthworm?

11. 'Animals depend entirely on plants.' Is this true? Give reasons.

12. Plants have not a nervous system like that of animals. Are they at all sensitive? Give reasons for your answer.

13. Describe the asexual reproduction of the amoeba.

14. Write a brief account of vegetative reproduction in plants.

15. Is asexual reproduction found in any animals except the amoeba?

16. Briefly describe what is meant by sexual reproduction.

17. What is meant by the terms (*a*) hermaphrodite, (*b*) unisexual?

18. Why is sexual reproduction the method of reproduction in nearly all living things?

19. Write an essay on 'Care of the young in animals'.

20. What must happen to an egg before it develops?

Chapter 10

1. Briefly describe how rocks have been formed.

2. What are fossils? How were they formed?

3. Write a few lines describing the world as you think it appeared millions of years ago.

4. How do modern horses differ from those of long ago?

5. Describe the ape-man. How did he differ from a man of today?

6. Do you find fossils in all rocks? Give reasons for your answer.

7. How do we know that man, mammals and flowering plants have only appeared recently on the earth?

8. What is meant by the following: (*a*) heredity, (*b*) the struggle for existence?

APPENDIX A

APPARATUS AND MATERIALS REQUIRED

The practical work in this book can be carried out with very little apparatus. Jam jars are extremely useful; and plain glass, screw-topped jars can be used for preserving specimens. Insect boxes can be bought or made (see Appendix B).

The following items are needed to study small specimens:

hand lenses
mounted needles
watch glasses
black and white tiles } one between two

1 doz. pipettes with rubber teats
1 microscope (or more) with two eye-pieces and 2 objectives
½ gr. microscope slides, 3 in. × 1 in.
1 doz. well slides
½ doz. coverslips, ½ in. diam. medium thickness.

To study fresh- and salt-water life it is necessary to have several aquaria. An electric aerator is useful but not necessary.

To preserve specimens the following items are required:

1 W.qt. formaldehyde, 40 %
4 doz. specimen tubes, 2 in. × 1 in.
1 doz. specimen tubes, 3 in. × 1 in.
1 doz. specimen tubes, 4 in. × 1 in.
8 oz. chloroform.

The following is a list of materials required to carry out the experimental work given in this book. The quantities given are those required for ONE group only. When possible the class should be divided into small groups, each with its own apparatus. The quantity of apparatus required will be determined by the number of groups.

1 Bunsen burner
1 tripod stand
1 iron-wire gauze square
1 batswing burner complete for bending glass-tubing (for whole class)
6 retort stands with boss and clamp
12 test-tubes, 5 in. × $\frac{3}{4}$ in.
2 hard-glass test-tubes, diam. 1 in., and corks with 1 hole to fit
2 small test-tubes, 2 in. × $\frac{3}{8}$ in.
1 test-tube stand to hold 8 test-tubes
1 test-tube holder
2 round flasks, flat-bottomed, 500 c.c.
2 corks to fit flasks, with (*a*) 3 holes, (*b*) 1 hole
4 triangular flasks, 500 c.c.
4 corks to fit, with 2 holes in each
4 beakers, 250 c.c.
1 soda-lime tube
6 measuring cylinders, 200 c.c.
2 measuring cylinders, 50 c.c.
1 dropping funnel, 50 c.c.
2 bell-jars large enough to hold a potted plant
2 square glass plates (thick glass), 10 in. × 10 in.
2 pieces of plain glass, 4 in. × 3 in.
8 large, wide-necked jars
8 shives to fit jars
2 filtering funnels, 5½ cm. diam.
2 aspirators each fitted with 2 bungs with 1 hole in each
6 hollow tubes, 8 in. × 1 in., lipped at one end
2 hollow tubes, 18 in. × 1 in., lipped at one end
1 Liebig condenser, 30 cm.
6 Petri-dishes
4 pipettes with rubber teats
1 pestle and mortar (medium size)
1 suction pump
1 packet of filter papers, 9 cm. diam.
1 book of blue litmus papers

1 book of red litmus papers
2 Mohr clips
2 rubber toy balloons
1 set of cork-borers
1 water bath (for whole class)
2 lb. of assorted glass tubing
¼ lb. each of 4 different widths of bore in capillary tubing
2 ft. of rubber tubing for Bunsen burner
4 ft. of rubber tubing, interior diam. 5 mm.
4 ft. of rubber tubing, interior diam. 3 mm.
1 B.D.H. soil-testing outfit including B.D.H. indicator
1 porcelain boat
1 capillary pipette
1 indicator chart
5 plant pots, 10 in. diam.
blotting paper
1 lb. tin of yellow vaseline
½ lb. roll of cotton wool
muslin
1 lb. starch
1 lb. glucose
2 lb. hydrochloric acid
1 pt. Fehling's A solution
1 pt. Fehling's B solution
½ lb. sodium bicarbonate
4 oz. iodine (1 % solution)
4 oz. litmus solution
1 pt. Millon's reagent
½ lb. caustic soda
½ pt. liquid paraffin or oil
1 pt. methylated spirits
½ lb. soda lime
1 lb. mercury
1 pt. cobalt chloride (5 % solution)
½ lb. potassium nitrate
½ lb. magnesium sulphate
½ lb. calcium sulphate
½ lb. acid sodium phosphate
½ lb. potassium chloride

½ lb. ferric phosphate
½ lb. ferric sulphate
½ lb. copper sulphate
½ lb. magnesium nitrate
½ lb. calcium nitrate
½ lb. sodium phosphate
½ lb. sodium nitrate
½ lb. potassium sulphate
1 oz. caustic potash pellets
4 oz. caustic potash (10 % solution)
2 oz. nutrient agar
1 oz. osmic acid
1 lb. each of large and small lead shot
100 c.c. essence of rennet
100 c.c. pepsin solution
4 oz. steapsin

Specimens

As far as possible children should find their own specimens. Names of dealers in specimens of livestock and microscope slides may be obtained from *The School Nature Study Union Journal*, which is issued in January, April, July and October. The annual subscription is five shillings and is due on 1 January. Subscriptions should be sent to:

THE HON. SUBSCRIPTION SECRETARY,
DR WINIFRED PAGE,
5 DARTMOUTH CHAMBERS,
THEOBALD'S ROAD, LONDON

APPENDIX B

SOME USEFUL HINTS

PART I

Siphoning

Place a jar of water above the aquarium. Take a piece of bent glass tubing with one arm reaching to the bottom of the jar. The other arm must be long enough to reach below the bottom of the jar on the outside. Suck the air out of the tube and the water will run freely. You can use this method to empty an aquarium.

The microscope

A microscope is an instrument used to make minute objects appear very much enlarged, so that these objects can be easily viewed with the eye. It consists of a combination of two lenses in a tube, one known as the OBJECTIVE and the other as the EYE-PIECE. An ordinary standard school-type microscope, as shown in Fig. 124, with two eye-pieces and two objectives is useful for elementary work. The microscope must be kept in a special case to keep it free from dust.

How to use the microscope

(*a*) Place the slide on the stage.

(*b*) Move the mirror until the light is reflected through the object.

(*c*) Move the *coarse* adjustment slowly, until you can see the object clearly.

(*d*) *High power objective.* After finding the object with the low power objective, turn round the nose-piece until the high power objective is under the draw-tube. Then bring the object into focus by carefully turning the *fine* adjustment.

N.B. Do not use the high power objective unless the object is covered with a coverslip.

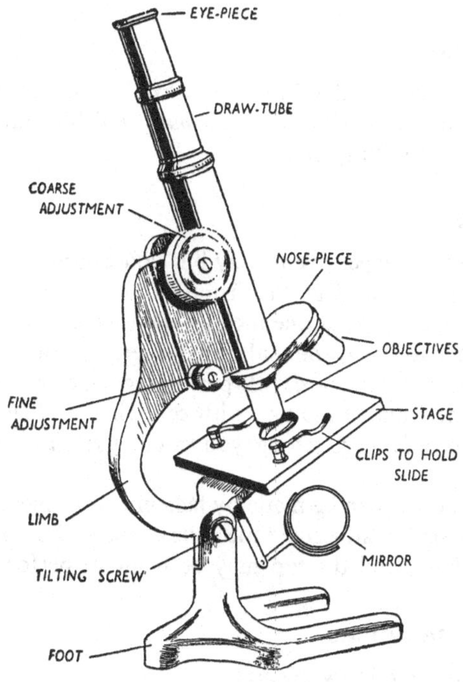

Fig. 124. *A microscope*

Microscope slides

Much time is necessary to acquire the technique of making permanent slides, and this is not a suitable type of work for the pupils who are not studying biology at an advanced level. It is, therefore, advisable to buy permanent slides. Temporary slides can be made in the following way:

1. Put your specimen on to the middle of a clean slide, and drop a spot of water on to it.

2. Put one edge of a clean coverslip on the edge of the drop of water, and lower the coverslip slowly, with the help of a mounted needle, being careful not to get an air bubble in the water.

Special well slides should be used if you are looking at small living specimens.

Trout hatchery

If you wish to see the development of fish in your school, obtain the pamphlet on *Trout Hatchery in School*, which is published by the School Nature Study Union.

Insect boxes

Insect boxes are expensive to buy but you can make your own in the following way. Cut the front out of a cardboard box, leaving an edging of about one inch all the way round. Make the back of the cage to open, and make a fastener. Cut out a rectangle and over this stitch a piece of perforated zinc. When this is done, cover the front of the box with cellophane. Then paint the box. If the box is very strong you could use glass instead of cellophane.

Glass insect cages can be made by fastening four pieces of glass together with strong adhesive tape. Plasticine will secure this to a cardboard base, and the top may be made of perforated zinc.

Preserving specimens

This may be done in two ways:

(*a*) *Wet.* Place the dead specimen in a tube or in a bottle of appropriate size containing a two per cent solution of formaldehyde.

(*b*) *Dry.* Small dead animals can be kept in insect boxes. These may be bought from dealers, or you can make wooden boxes with glass lids, or you can make them with match boxes or tins that have no lids. These can be covered with cellophane.

As soon as the animal is dead it should be spread out into the required position before it becomes stiff. You may either rest the specimen on cotton wool inside the box, or you can stick a pin through the thorax of the animal. If pins are used, you must put a layer of cork on the bottom of the box, into which the pin may be stuck. Place a little naphtha, or moth balls or a similar substance in the boxes to preserve the animals.

To kill animals

If you wish to preserve animals that are not dead, kill them first by putting them into a killing bottle. Special killing bottles can be bought from dealers or you can make your own in the following way. Put some cotton wool that has been soaked in chloroform in the bottom of a wide-necked bottle. Then place some dry cotton wool on top of this.

Specimens

In addition to the specimens mentioned in the book, the following may be found useful:

Reptiles. Nearly all English reptiles can be kept in school in a suitable vivarium (see Book I, Fig. 63).

Dead specimens may be pickled in five per cent formaldehyde, and are very useful when living specimens are not available. Children could try to find sloughed skins of snakes.

Birds. Many useful specimens may be collected by the children and brought to school. For instance: (*a*) feathers of all kinds; (*b*) skeletons of birds whose flesh has been eaten; (*c*) skeletons of dead birds that have been buried in the ground; (*d*) teachers must give advice on the advisability of collecting birds' eggs and nests.

Mammals. A very wide range of specimens may be brought by the children. Dead specimens of small mammals or baby animals may be found useful if pickled in five per cent formaldehyde. Skulls, hoofs and bones of all kinds will be collected by children. Do not keep living mammals unless you have suitable living accommodation for them.

Sea water

Ingredients for making artificial sea water are given in *Introduction to Zoology through Nature Study*, by R. Lulham.

Nature study records

The keeping of nature study records makes pupils more observant and gives them practice in recording accurately what they have seen.

NATURE STUDY CALENDARS. Pupils should begin their calendars in January and record dates, times and places where specimens are seen. In this way they will have an accurate account of the seasonal activities of living things. The effect of climate on these activities will be seen if records are kept for several years together with observations on weather.

FLOWER RECORDS. When collecting wild flowers, pupils should note when and where they are found. They will then know the flowering season of each plant as well as its habitat.

INDIVIDUAL RECORDS can be made throughout a year, of any one specimen chosen by the boy or girl. For instance, in the study of a tree, the following things must be noted: its appearance at each season; the dates when buds open, flowers appear, fruits are formed and are scattered and leaves fall. Pupils should make records of all animals that they keep, and, when possible, include stages in development from egg to adult. Diagrams could be used to illustrate this work.

PART II

Growth of bacteria

Bacteria can be grown in Petri-dishes or in test-tubes, in the following way:

1. Sterilize the Petri-dishes by putting them in the domestic oven (temp. 160° C.) for half an hour.
2. Put nutrient agar into dish and let it set. (*N.B.*—Petri-dish must be covered.)
3. Infect your agar in any of the following ways:

    (*a*) Expose agar to air for five minutes.
    (*b*) Put a drop of milk on to the agar.
    (*c*) Put a drop of boiled milk on to the agar.

(*d*) Allow a fly to walk over the agar (keeping lid on whilst he is inside).
(*e*) Put dust on to agar.
(*f*) Place 3 hairs on to agar.
(*g*) Put spot of dirty water on to agar.

Water culture solutions

(*a*) Complete:

2 litres distilled water
2 gm. potassium nitrate
$\frac{1}{2}$ gm. magnesium sulphate
$\frac{1}{2}$ gm. calcium sulphate
$\frac{1}{2}$ gm. ferric phosphate

(*b*) No nitrogen. Use potassium sulphate instead of potassium nitrate.

(*c*) No potassium. Use sodium nitrate instead of potassium nitrate.

(*d*) No phosphorus. Use ferric sulphate instead of ferric phosphate.

(*e*) No calcium. Use magnesium nitrate instead of calcium sulphate.

(*f*) No sulphur. Use magnesium nitrate and calcium nitrate instead of magnesium sulphate and calcium sulphate.

(*g*) No iron. Use sodium phosphate instead of ferric phosphate.

Enzyme experiments

EXPERIMENT TO SHOW THE ACTION OF PTYALIN ON STARCH. Place a little starch solution into each of two test-tubes and add saliva to one. Leave the test-tubes in a water bath at body temperature for five minutes. Test both test-tubes for glucose (see page 142).

EXPERIMENT TO SHOW THE ACTION OF PEPSIN ON PROTEINS. Place a little powdered egg and water into two test-tubes (or use soluble proteins such as fibrin in blood). Add pepsin solution to one test-tube, and leave both at body temperature for some time. (Pepsin solution can be made by dissolving 0·25 gm. of pepsin powder in 25 c.c. of water and then adding an equal volume of

0·4 per cent hydrochloric acid.) Add a little caustic soda to each test-tube and then one drop of copper sulphate. Proteins show a purple coloration, whereas peptones give a rose colour.

EXPERIMENT TO SHOW THE ACTION OF STEAPSIN ON FATS. Pour a little milk into each of two test-tubes, and add litmus solution to each. Add steapsin to one test-tube. Keep both test-tubes at body temperature for a short time. The litmus turns pink in the test-tube containing steapsin. This reaction is brought about by the fatty acid which is formed when fats are broken down.

EXPERIMENT TO SHOW THE ACTION OF RENNIN ON MILK. Pour a little milk into two test-tubes, and add essence of rennet to one. Leave in a water bath at body temperature for a short time. The milk in the test-tube containing rennin is coagulated.

To dilute acids. Acids are diluted when they are mixed with water, BUT ACID SHOULD ALWAYS BE POURED INTO THE WATER SLOWLY, *not* water into the acid.

APPENDIX C

The following books should be included in the School science library, and used by the children to identify plants and animals that they find which are not mentioned in this book.

BATTEN, H. MORTIMER. *Our Garden Birds.*
BENTHAM AND HOOKER. *A British Flora.* 2 vols.
BIBBY, C. *How Life is Handed on.*
BOULENGER, E. G. *A Naturalist at the Zoo.*
BOULENGER, E. G. *Zoo Cavalcade.*
BOULENGER, E. G. *World Natural History.*
BOULENGER, E. G. *The Aquarium.*
BRIMBLE, L. J. F. *Intermediate Botany.* 3rd ed.
BRIMBLE, L. J. F. *Trees in Britain.*
BUCHSBAUM, R. *Animals without Backbones.*
COWARD, T. A. *Life of the Wayside and Woodland.*
COWARD, T. A. *Birds of the British Isles and their Eggs.* 3 vols.
DAYLISH, E. TITCH. *Name this Bird.*
DUNCAN, E. MARTIN (ed.). *Cassell's Natural History.*
ELMHURST, R. *Naturalist at the Sea Shore.*
FURNEAUX, W. *Life in Ponds and Streams.*
FURNEAUX, W. *The Sea Shore.*
FURNEAUX, W. *Human Physiology.*
GROOM, PERCY. *Elementary Botany.*
HELLYER, A. G. L. *Garden Pest Control.*
HUXLEY, T. H. *Lessons in Elementary Physiology.*
IMMS, A. D. *Insect Natural History.*
JENKINS, J. T. *Fishes of the British Isles.*
JOHNS, Rev. C. A. *Flowers of the Field.*
JOY, NORMAN H. *British Beetles. Their Homes and Habits.*
KELMAN, J. H. *Butterflies and Moths.*
KELMAN, J. H. *Bees.*

LULHAM, R. *An Introduction to Zoology through Nature Study.*
MELLANBY, H. *Animal Life in Fresh Water.*
POCHIN, E. *How to Recognize Trees of the Countryside.*
RAMSBOTTOM, J. *Mushrooms and Toadstools.*
ROLFE, R. T. and F. W. *Romance of the Fungus World.*
ROMER, A. S. *Man and the Vertebrates.*
SIDEBOTHAM, H. *Wild Animals.*
SOUTH, R. *Moths of the British Isles.* Volumes I and II.
STEP, EDWARD. *British Insect Life.*
STEP, EDWARD. *Bees, Wasps, Ants and Allied Insects of the British Isles.*
STEP, EDWARD. *Toadstools and Mushrooms of the Countryside.*
STEP, EDWARD. *Wild Flowers Month by Month in their Natural Surroundings.* 2 vols.
STEP, EDWARD. *Wayside and Woodland Blossoms.* Series I, II and III.
THOMPSON, H. S. *How to Collect and Dry Flowering Plants.*
TINN, FRANK. *Eggs and Nests of British Birds.*
WATTS, W. M. *A School Flora.*
WELLS, H. G. and G. P. and HUXLEY, J. *Evolution.*
PELICAN BOOKS:
  WOKES, F. *Food, the Deciding Factor.*
PUFFIN BOOKS:
  Animals of the Countryside.
  Trees in Britain.
  The Story of Plant Life.
  A Book of Insects.
  Butterflies in Britain.
'OBSERVER'S' BOOKS:
  British Wild Animals.
  British Butterflies.
  British Wild Flowers.
  Trees and Shrubs of the British Isles.
'SHOWN TO THE CHILDREN' BOOKS:
  BLACKIE, A. H. *Nests and Eggs.*
  KELMAN, J. H. *Flowers.*
  KELMAN, J. H. *The Sea Shore.*
  SCOTT, M. K. *Birds.*
  WYSS, C. VON. *Living Creatures.*

'HOW TO IDENTIFY' BOOKS:

British Butterflies.

British Birds.

British Wild Animals.

British Wild Flowers.

Trees and Shrubs of the British Isles.

'HOW TO RECOGNIZE' BOOKS:

POCHIN, E. *British Wild Birds.*

POCHIN, E. *British Birds, Eggs and Nests.*

'FOR THE POCKET' BOOKS:

SANDARS, EDMUND. *A Butterfly Book.*

SANDARS, EDMUND. *A Beast Book.*

SANDARS, EDMUND. *A Bird Book.*

POCKET GUIDE SERIES:

Wild Flowers of the Wayside and Woodland. Compiled by T. N. SCOTT and W. J. STOKOE.

Birds of the Wayside and Woodland. T. A. COWARD.

Butterflies and Moths of the Wayside and Woodland. Compiled by W. J. STOKOE.

YOUNG FARMERS' CLUB BOOKLETS:

1. *The Farm.* 2. *Rabbit Keeping.* 3. *Pig Keeping.* 4. *Poultry Keeping.* 5. *Bees.* 6. *Goat Keeping.* 9. *Garden and Farm Pests.* 10. *Cows and Milk.* 11. *Ducks, Geese and Turkeys.* 13. *Farm Horses.* 16. *Sheep Farming.*

Lists of very useful pamphlets, bulletins, booklets and pictures of plants and animals may be obtained from:

1. The Natural History Museum, South Kensington, London.
2. The Ministry of Agriculture and Fisheries.

INDEX

Printed in the United States
By Bookmasters